INSTITUTE OF GEOLOGICAL SCIENCES

Natural Environment Research Council

Mineral Assessment Report 34

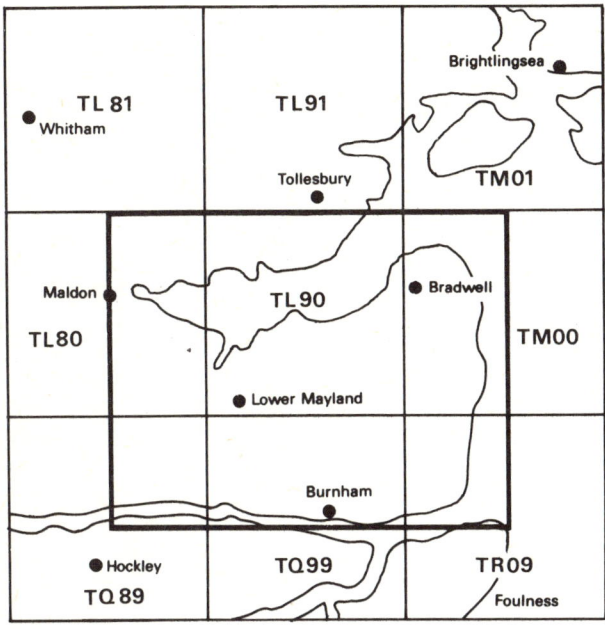

The sand and gravel resources of the Dengie Peninsula, Essex

Description of 1:25 000 sheet TL 90 and parts of sheets TL 80, TM 00, TQ 89, TQ 99 and TR 09

M. B. Simmons

ISBN 0 11 884081 9

London   Her Majesty's Stationery Office   1978

The first twelve reports on the assessment of British sand and gravel resources appeared in the Report Series of the Institute of Geological Sciences as a subseries. Report No. 13 onwards appear in the Mineral Assessment Report Series of the Institute. Details of published reports appear at the end of this report.

Any enquiries concerning this report may be addressed to Head, Industrial Minerals Assessment Unit, Institute of Geological Sciences, Keyworth, Nottingham NG12 5GG

The asterisk on the front cover indicates that parts of sheets adjacent to that quoted are described in this report.

PREFACE

National resources of many industrial minerals may seem so large that stocktaking appears unnecessary, but the demand for minerals and for land for all purposes is intensifying and it has become increasingly clear in recent years that regional assessments of the resources of these minerals should be undertaken. The publication of information about the quantity and quality of deposits over large areas is intended to provide a comprehensive factual background against which planning decisions can be made.

Sand and gravel, considered together as naturally occurring aggregate, was selected as the bulk material demanding the most urgent attention, initially in the south-east of England, where about half the national output is won and very few sources of alternative aggregates are available. Following a short feasibility project, initiated in 1966 by the Ministry of Land and Natural Resources, the Industrial Minerals Assessment Unit (formerly the Mineral Assessment Unit) began systematic surveys in 1968. The work is now being financed by the Department of the Environment and is being undertaken with the co-operation of the Sand and Gravel Association of Great Britain.

This Report describes the resources of sand and gravel of 182.4 km^2 of country between the estuaries of the River Crouch and River Blackwater, shown on the accompanying 1:25 000 resource sheet. The survey was conducted by Mr S.E. Hollyer and Ms M.B. Simmons in collaboration with Mr R.A. Ellison of the East Anglia and South-East England Field Unit and Mr M. Sarginson of the Engineering Geology Unit during 1973 and 1974. The work is based on a geological survey at 1:10 650 carried out in 1966 and 1969 by Dr C.R. Bristow and in 1972 to 1973 by Dr M.R. Henson: other members of the East Anglia and South-East England Field Unit, particularly Mr R.D. Lake, offered advice and Mr Sarginson provided information upon which Appendix H is based.

Mr J.W. Gardner, CBE (Land Agent) has been responsible for negotiating access to land for drilling. The ready co-operation of landowners and tenants in this work is gratefully acknowledged.

Austin W. Woodland
Director

Institute of Geological Sciences
Exhibition Road
South Kensington
London SW7 2DE
24 November 1978

ii

CONTENTS

Summary   1

Introduction   1

Description of the Dengie Peninsula   3

  General   3
  Topography   3
  Geology   3
  Composition of the sand and gravel   9
  The map   10
  Results   10
  Notes on resource blocks   14

Appendix A: Field and laboratory
procedures   16

Appendix B: Statistical procedure   16

Appendix C: Classification and
description of sand and gravel   17

Appendix D: Explanation of
borehole records   21

Appendix E: Boreholes used in the
assessment of resources   23

Appendix F: Summary of information from
Industrial Minerals Assessment Unit
and Engineering Geology Unit boreholes   25

Appendix G: Industrial Minerals Assessment
Unit borehole records and exposure
records   28

Appendix H: Resistivity survey results   85

Appendix J: List of quarries on the
Dengie Peninsula   87

Appendix K: Conversion table - metres
to feet   88

References   89

FIGURES

1   Sketch-map showing the location of the
    resource sheet (Dengie Peninsula) with
    resource block boundaries   2

2   Geological sketch-sections showing the
    sequence of deposits in the north and south
    of the Dengie Peninsula   4

3   The form of the surface of the bedrock
    (London Clay and Claygate Beds), shown
    by contours plotted from 160 sample
    points   7

4   Regional grading characteristics of the
    mineral based on 29 mineral assessment
    boreholes and five exposures   12

5   Particle-size distribution for the
    assessed thickness of mineral in
    resource blocks A, B and C   13

6   Example of resource block assessment:
    calculation and results   19

7   Example of resource block assessment:
    map of fictitious block   20

8   Diagram to show the descriptive
    categories used in the classification of
    sand and gravel   20

9   Terrameter traverse in the region of
    the Burnham Buried Channel   85

10   Section across the Burnham Buried
     Channel   85

11   Apparent resistivity ranges for
     certain lithologies and superficial
     deposits in south-east Essex   86

Map   The sand and gravel resources of
      the Dengie Peninsula in pocket

TABLES

1   Geological succession in the Dengie
    Peninsula   3

2   Results of 10 per cent fines, specific
    gravity and water absorption tests   10

3   The sand and gravel resources of
    the Dengie Peninsula   11

4   Classification of gravel, sand and
    fines   18

5   Results of expanding traverse near
    Dammer Wick Farm   86

The sand and gravel resources of the Dengie Peninsula, Essex

Description of 1:25 000 sheet TL 90 and parts of sheets TL 80, TM 00, TQ 89, TQ 99 and TR 09

M. B. Simmons

SUMMARY

The geological maps of the Institute of Geological Sciences, pre-existing borehole information, 59 boreholes drilled for the Industrial Minerals Assessment Unit and 8 for the South-East England Field Unit and the Engineering Geology Unit, form the basis of the assessment of sand and gravel resources on the Dengie Peninsula, Essex. Additionally, use was made of information obtained from Dutch probes, Delft boreholes and resistivity work carried out by the Engineering Geology Unit.

All deposits in the area which might be potentially workable for sand and gravel have been investigated and a simple statistical method has been used to estimate the volume. The reliability of the volume estimates is given at the symmetrical 95 per cent probability level.

The 1:25 000 map is divided into three resource blocks, containing between 12.3 and 13.9 km^2 of mineral-bearing ground. For each block the geology of the deposits is described, and the mineral-bearing area, the mean thicknesses of overburden and mineral and the mean gradings are stated. Detailed borehole data are also given. The geology and topography, the positions of the boreholes and exposures and the outlines of the resource blocks are shown on the accompanying map.

Bibliographic reference

SIMMONS, M.B. 1978. The sand and gravel resources of the Dengie Peninsula, Essex. Description of 1:25 000 sheet TL 90 and parts of sheets TL 80, TM 00, TQ 89, TQ 99 and TR 09. Miner. Assess. Rep. Inst. Geol. Sci., No. 34.

Author

M.B. Simmons, B.Sc.
Institute of Geological Sciences, London.

INTRODUCTION

The survey is concerned with the estimation of resources, which include deposits that are not currently exploitable but have a foreseeable use, rather than reserves, which can only be assessed in the light of current, locally prevailing, economic considerations. Clearly, both the economic and the social factors used to decide whether a deposit may be workable in the future cannot be predicted; they are likely to change with time. Deposits not currently economically workable may be exploited as demand increases, as higher grade or alternative materials become scarce, or as improved processing techniques are applied to them. The improved knowledge of the main physical properties of the resource and their variability, which this survey seeks to provide, will add significantly to the factual background against which planning policies can be decided (Archer, 1969; Thurrell, 1971; Harris and others, 1974).

The survey provides information at the 'indicated' level "for which tonnage and grade are computed partly from specific measurements, samples or production data and partly from projection for a reasonable distance on geological evidence. The sites available for inspection, measurement, and sampling are too widely spaced to permit the mineral bodies to be outlined completely or the grade established throughout". (Bureau of Mines and Geological Survey, 1948, p. 15).

It follows that the whereabouts of reserves must still be established and their size and quality proved by the customary detailed exploration and evaluation undertaken by the industry. However, the information provided by this survey should assist in the selection of the best targets for such further work.

The following arbitrary physical criteria have been adopted.

a. the deposit should average at least one metre in thickness,

b. the ratio of overburden to sand and gravel should be no more than 3:1,

c. the proportion of fines (particles passing the No. 240 mesh BS sieve, about 1/16 mm) should not exceed 40 per cent,

d. the deposit must lie within 25 m of the surface, this being taken as the likely maximum working depth under most circumstances. It follows from the second criterion that boreholes are drilled no deeper than

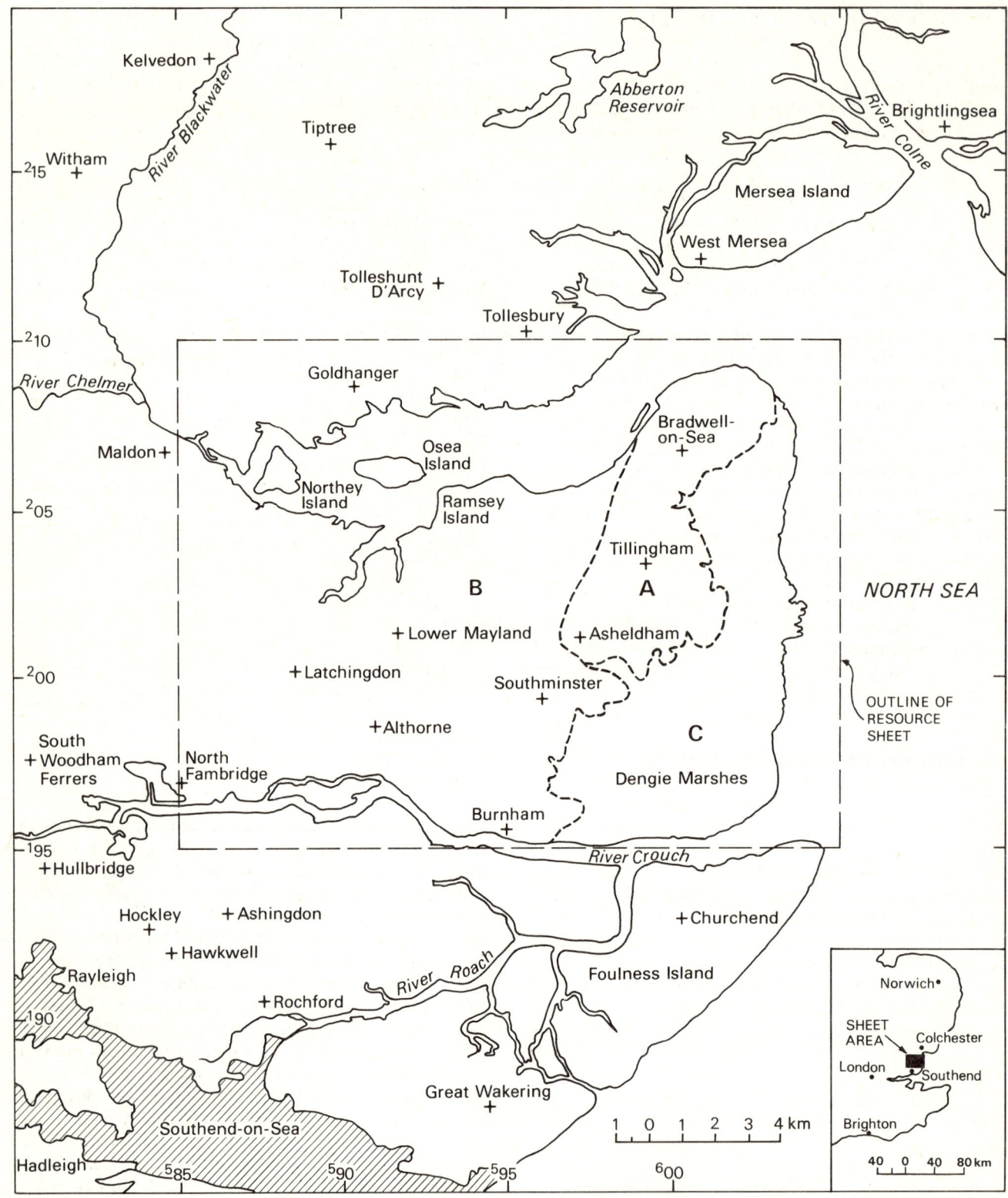

Fig. 1   Sketch-map showing the location of the resource sheet (Dengie Peninsula)
with resource block boundaries

18 m if no sand and gravel has been proved.

A deposit of sand and gravel which broadly meets these criteria, is regarded as 'potentially workable' and is described and assessed as 'mineral' in this report. As the assessment is at the indicated level, parts of such a deposit may not satisfy all the criteria.

For the particular needs of assessing sand and gravel resources, a grain-size classification based on the geometric scale 1/16 mm, 1/4 mm, 1 mm, 4 mm, 16 mm has been adopted. The boundaries between fines (that is, the clay and silt fractions) and sand, and between sand and gravel grade material, are placed at 1/16 mm and 4 mm respectively (see Appendix C).

The volume and other characteristics are assessed within resource blocks, each of which, ideally, contains approximately 10 km^2 of sand and gravel. No account is taken of any factors, for example, roads, villages and high agricultural or landscape value, which might stand in the way of sand and gravel being exploited, although towns are excluded. The estimated total volume therefore bears no simple relationship to the amount that could be extracted in practice.

It must be emphasised that the assessment applies to the resource block as a whole. Valid conclusions cannot be drawn about the mineral in parts of a block, except in the immediate vicinity of the actual sample points.

DESCRIPTION OF THE DENGIE PENINSULA

GENERAL

The survey area includes 182.4 km^2 of the generally rather flat-lying land of the Dengie Peninsula, stretching from the River Crouch in the south to the River Blackwater in the north (including Osea Island) and inland as far west as grid line 85 E. It is characterised by agricultural development, with scattered communities including Southminster, Tillingham and Bradwell-on-Sea. The holiday resort of Burnham-on-Crouch (population 5000) is the largest town in the area, although part of Maldon (population 14 000) lies within the limits of the survey. Those parts of TL 80 and TL 90 north of the Blackwater have been previously assessed and the results of the surveys are included in published reports (Ambrose 1973a, Ambrose 1973b).

TOPOGRAPHY

The reclaimed alluvial areas of Bradwell, Tillingham and Dengie Marshes, flanking the coast, form extensive flats, rarely rising more than 2.5 m above Ordnance Datum. The remainder of the peninsula is gently undulating with occasional relatively steep slopes, particularly near the northern bank of the Crouch and to the south of the Blackwater Estuary in the vicinity of Steeple and St Lawrence. The maximum recorded height in the area, 51.5 m (169 ft), is attained to

the west of Althorne. The area is drained by a number of minor streams which flow eastwards from a low ridge that runs south-south-east from Tillingham. The streams flow across the marshes and into the North Sea.

GEOLOGY

The deposits that crop out on the Dengie Peninsula are shown in Table 1, which is followed by an account of the drift and solid strata, including a description of their lithology and occurrence, a discussion on the sub-Drift surface and the interrelationships of the deposits.

Table 1 Geological succession in the Dengie Peninsula

DRIFT-Recent and Pleistocene
River Alluvium, Storm Gravel Beach Deposits, Marine Beach and Tidal-Flat Deposits (present day) and Marine or Estuarine Alluvium including Beach Deposits
River Terrace and Fluvio-glacial Sand and Gravel (including kame deposits) and River Loam (Brickearth)
Buried Channel Deposits
Head (including Head Brickearth)
Glacial Sand and Gravel
SOLID-Eocene
Claygate Beds
London Clay

SOLID

London Clay

The London Clay is the oldest formation exposed, and the oldest proved in boreholes in this survey. Earlier, deeper boreholes for water, however, penetrated the Lower London Tertiaries (which comprise the Thanet Beds, the Woolwich Beds and the Oldhaven Beds) and the Chalk (see Hydrogeological Department record, 242/23 (Davies, M.C. and others, 1965)). When fresh, the London Clay in this area is stiff, dark olive-grey, slightly silty clay, but this was found in only 17 boreholes of the 67 drilled in this survey: three of the 17 penetrated at least 6 m of Claygate Beds before passing into the underlying London Clay. Eight boreholes proved unweathered London Clay at less than 1 m below the base of the terrace deposits, but at each there was evidence that localised channelling had cut deeper into the underlying bedrock than the average basal bench level of the terrace. Beneath the Marine or Estuarine Alluvium the depth of weathering was often in excess of 4 m. Near the ground surface the London Clay usually weathers dark yellowish brown, with blue reduction veins along fissure planes. Selenite is commonly found in the weathered clay.

Claygate Beds

There is a gradual transition from the London Clay to the overlying Claygate Beds, the youngest solid formation in the Dengie area. Recent investigations by the East Anglia and South-East England Field Unit have suggested that the base

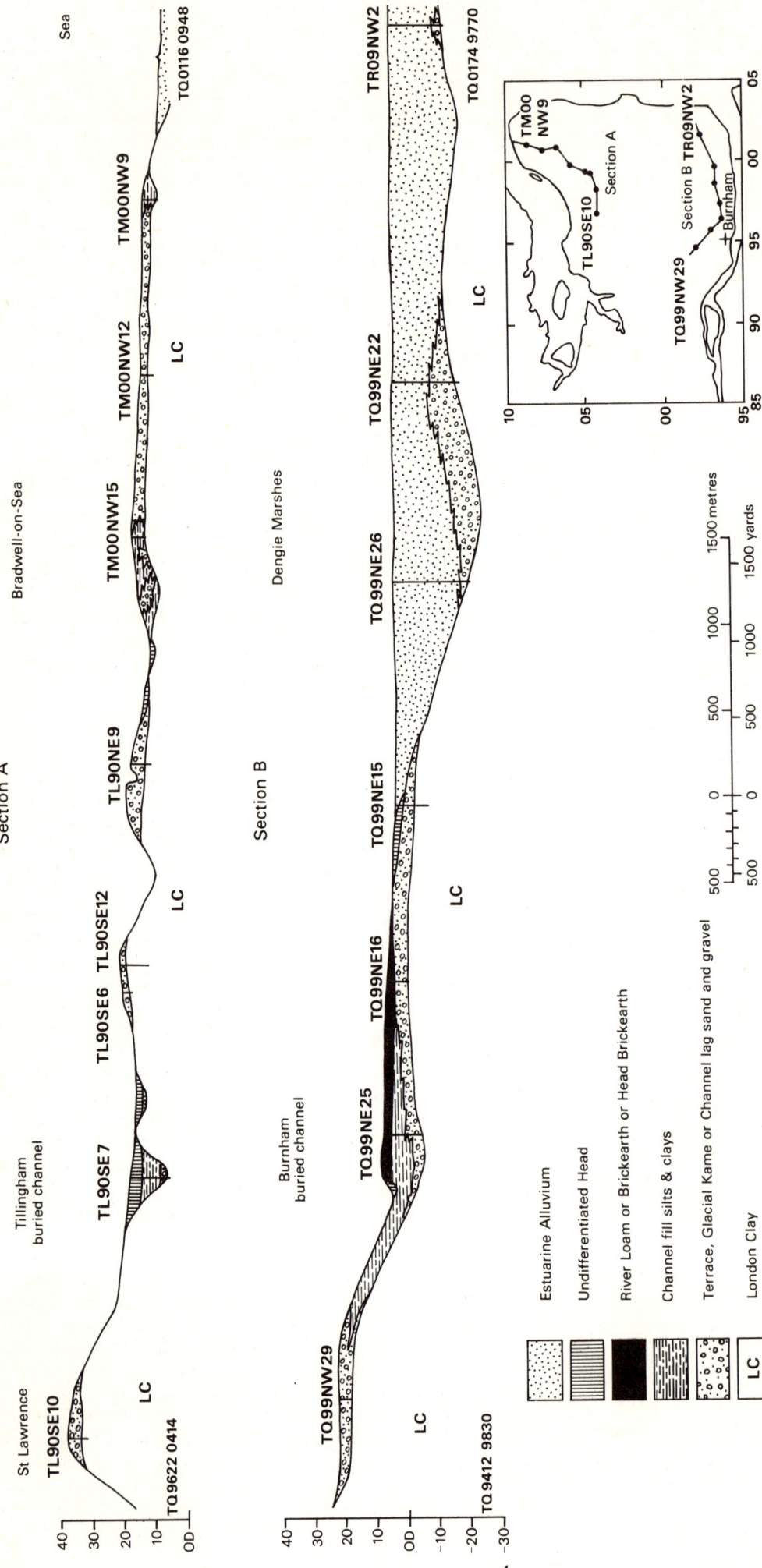

Fig. 2 Geological sketch-sections showing the sequence of deposits in the north and south of the Dengie Peninsula

4

of the Claygate Beds should be taken at the lowest recognisable major sandy horizon. Although this criterion may be used conveniently in deep boreholes, it is more difficult to apply in field mapping.

The Claygate Beds consist of interbedded silty clays and sandy silts with occasional shells, and the clays are very similar in appearance to London Clay. They are restricted to the ground to the west of Burnham [950 960], where they cap the high ground above about 30 m above Ordnance Datum. The maximum thickness proved by boreholes was 12 m.

DRIFT

Glacial Sand and Gravel
The only occurrence [854 063] of Glacial Sand and Gravel shown on the resource map, in the extreme north-west of the area, has not been sampled as it lies within the urban area of Maldon. However, other deposits of possible glacial origin, mapped as terrace gravels, are included in the main mass of sand and gravel deposits between Burnham and Bradwell-on-Sea. These are discussed in the section on River Terrace Sand and Gravel.

Head (including Head Brickearth)
The term 'Head' covers deposits of widely differing ages, formed by solifluction and hill creep, the local lithology depending upon the parent material. The most common head material was derived from the London Clay. and is usually firm to stiff silty clay, yellowish brown in colour, with occasional 'race' (calcium carbonate concretions up to 10 mm in diameter), carbonaceous material and flint pebbles. It generally averages about 1 m in thickness and occurs in the valleys and mantling slopes. Head occasionally underlies sand and gravel, as in boreholes TL 90 SE 9 and TQ 99 NW 29, but may post-date them elsewhere, as in borehole TQ 99 NE 16, where material derived from older river terrace deposits covers the First Terrace Sand and Gravel. In borehole TQ 99 NE 20, Head, which overlies channel deposits and was probably derived from the higher gravels to the west, was found to be of mineral grade.

Deposits to the east of Burnham centred at [960 965] shown on the resource map as Head (Undifferentiated) were mapped at the 1:10 560 scale as Head and Head Brickearth. Although the latter deposit resembles River Loam (see below) lithologically, being orange clayey sandy silts with occasional pebble stringers, much of the material has been affected by solifluction, particularly where it mantles the slope between the First and Third terraces. These deposits are distinguished from River Loam, therefore, by their mode of occurrence, either on or at the base of slopes and proved, for example, in boreholes TQ 99 NE 25 and TQ 99 NE 16, and from Head by their relative lithological uniformity. These criteria suggest that the Head Brickearth originated in part possibly as loess, a windblown, fine-grained blanket deposit, which was subsequently subjected to solifluction processes during a periglacial period.

Buried Channel Deposits
A small patch of Older Estuarine Alluvium forms a slight feature above the Marine or Estuarine Alluvium to the east of Southminster. This consists of firm to stiff pale grey clay and silt with 'race' and rootlets and occasional marine or estuarine shells. Similar material was augered to the south-east of Southminster near Rumbolds [9668 9842] and to the north-east of Burnham near Brook Farm [9586 9729]. Borehole TQ 99 NE 25 penetrated 3.7 m of pale grey clay overlying gravel, and borehole TQ 99 NE 20 near Goldsand Bridges proved at least 1.3 m of channel deposits, which were not bottomed. This is the Burnham Buried Channel. In the vicinity of this borehole a resistivity survey, consisting of four expanding traverses and a constant separation traverse perpendicular to the expanding traverses, indicated a channel feature trending north-east to south-west, whose deepest part is at least 5 m below Ordnance Datum (cross-section Fig. 10, Appendix H). It is likely that at least part of the Older Estuarine Alluvium cropping out to the north was deposited within this channel and originally extended southward but was removed by more recent marine erosion prior to the deposition of the Marine or Estuarine Alluvium.

Clays overlying the London Clay in boreholes to the east of the Older Estuarine Alluvium, for example, in borehole TQ 99 NE 23, are lithologically similar and may possibly be of the same age as the channel deposits mentioned above. From this evidence, and from the map of the London Clay surface, it is suggested that this channel may run to the north-east from Burnham to Southminster, whence it turns to the east and the sea.

The channel shape and dimensions suggested by the geophysical evidence, however, (see Fig. 10) possibly indicate two phases of channel development. An earlier, shallower channel, which may have been linked with the Rochford Channel to the south (Hollyer, 1978), was later overdeepened by a second channel, which removed a section of the former channel-fill sequence and eventually silted up with alluvial silts and clays.

Clays containing an organic-rich horizon are exposed near Ratsborough [950 985]. From their position between the overlying Third Terrace and the pebbly sands of possible fluvio-glacial origin, they are thought to be channel-fill deposits of the older channel

The deposits infilling the narrow, steep-sided channel near Bradwell-on-Sea (see below) are rich in organic matter, generally more coarse grained and more variable in lithology than the sediments of the later Burnham Channel proved in borehole TQ 99 NE 25, and may be of approximately the same age as the silts and clays of the older Burnham Channel. They filled a depression before the Third Terrace gravels were laid down as proved by borehole TL 90 NE 7, in which a thin bed of gravel, presumed to be a remnant of the Third Terrace Deposits, was found to overlie the channel-fill material.

The origin of the fine-grained deposits is uncertain, for although the steep-sided, narrow profiles of the Bradwell Channel and of the Rochford Channel point to erosion and infilling in permafrost conditions, the absence of recognisable till and the fairly high organic content indicating local plant colonisation, suggest deposition in a temperate climate. It is likely, therefore, that these channels had a fairly complex history with several phases of erosion and infilling with much reworking of the sediments (Lake and others, 1977).

Beneath the Estuarine Alluvium, irregular and impersistent beds of gravel were found to overlie the London Clay in some boreholes. The gravels are confined almost exclusively to the area to the south of Dengie village, being found most consistently within about two kilometres of the south coast of the Peninsula. There is considerable variation in thickness of the gravels, particularly when they infill the deeply incised channels cutting into the London Clay surface. It has been deduced from the form of the channels that these gravels are fluvial or fluvio-glacial in origin, having been deposited at times of lowered sea level during the Pleistocene.

The gravels in the vicinity of the Crouch were probably deposited by that river when it was at or near its lowest level. Fossils from near the base of the deepest channel are thought to be pre-Flandrian in age although, from their appearance, it is probable that they have been reworked. There may have been some reworking of the upper part of these deposits as the sea transgressed across them, as evidenced by Cardium and other marine shells often found in the upper few feet, and they were redeposited. Similar deposits occur today beyond the sea wall, particularly in the Bradwell area at Sales Point [030 085]. In addition to the gravels found overlying the London Clay, similar deposits were occasionally encountered within the Alluvium, probably formed by reworking of the earlier gravels during a period of marine transgression.

River Terrace and Fluvio-glacial Sand and Gravel (including kame deposits)

The main mass of sand and gravel overlies the London Clay and occasionally the Claygate Beds in a NNE–SSW-trending belt running parallel to the landward edge of the Marine or Estuarine Alluvium marshes to the east. In addition, two small patches of terrace deposits totalling not more than 1 km^2 in area, overlie the London Clay on Osea Island and Ramsey Island [946 056] in the Blackwater Estuary.

Geological mapping suggested that these deposits could be correlated with the First, Second, Third and Fourth terraces of the Southend Peninsula to the south of the River Crouch, (Hollyer, 1978) owing to their terrace-like morphology. However, evidence from recent exposures in working sand and gravel pits and from the boreholes drilled in this assessment programme, suggests a more complex origin for most.

The much-dissected patches of deposits mapped as Fourth Terrace on the higher slopes of the ridge of London Clay or Claygate Beds, which trends NNE–SSW through Tillingham and Althorne have base levels ranging from less than 30 m to nearly 50 m above Ordnance Datum. These high-level gravels were found to be very clayey and silty with never more than 20 per cent pebbles, were occasionally overlain by organic-rich, cryoturbated, silty clays, and may be more directly glacial or fluvio-glacial than strictly fluvial in origin.

Many boreholes and some pit exposures in the deposits mapped as Third Terrace Sand and Gravel, which form more than half of the exposed mineral in this area, proved lithological sequences that suggested that fluvial processes alone were not responsible for the deposition of this material or of the Second Terrace deposits around Bradwell. 'Clayey' or 'very clayey' sands occasionally underlain by a coarse lag deposit and sometimes overlain by more typical terrace gravels were frequently encountered. These pale, silty, pebbly sands are exposed in several pits, where they were seen to be trough cross-bedded sands. Gravel is almost entirely confined to the surface of the foresets, as for example, at the pits [9849 0301], near Stow's Farm, Tillingham, near Ratsborough, Southminster at [9512 9851] and at [9594 9931] near Goldsand Road, Southminster.

Lateral variations seen at the last-named exposure and in a recent south-eastward extension of the Ratsborough pit [950 985] strengthen the theory that the sandy facies of these deposits are not terrace deposits, although they were deposited on a typical terrace-like bench feature at between 10 m and 19 m above Ordnance Datum. At Goldsand Road pit, trough cross-bedded gravelly sands pass laterally eastwards into micro-trough cross-laminated silts with abundant roots. Towards the south-east of the Ratsborough pit, cryoturbated sandy gravel overlies laminated clays containing an organic horizon and this argillaceous facies in turn overlies pebbly sand. Comparable lithologies were encountered in the deposits mapped as Second Terrace around Bradwell but no exposures were seen.

These features can best be explained if glacial influence is invoked, and it is possible that an ice lobe advanced from the North Sea as far as the Tillingham-Althorne ridge of London Clay and Claygate Beds. At the time of maximum advance, kames ('Fourth Terrace' deposits) were built up at the ice front, while subglacial streams eroded into the bedrock producing steepsided channels parallel with the ice front. One such channel trending NNE-SSW can be identified on the subdrift contour map to the south-west of Bradwell-on-Sea (Fig. 3).

It is thought that glacial outwash streams emanating from the waning ice sheet deposited silty pebbly sands on the irregular post-glacial topography. Later deposition of laminated grey silts and clays with organic debris infilled the channel depressions; these deposits are discussed in the section on Buried Channel Deposits (p. 5).

Following the infilling of the channels, at a time when the Third Terrace gravels of the Rochford area to the south were being deposited, it is thought that a floodplain of a proto-Thames

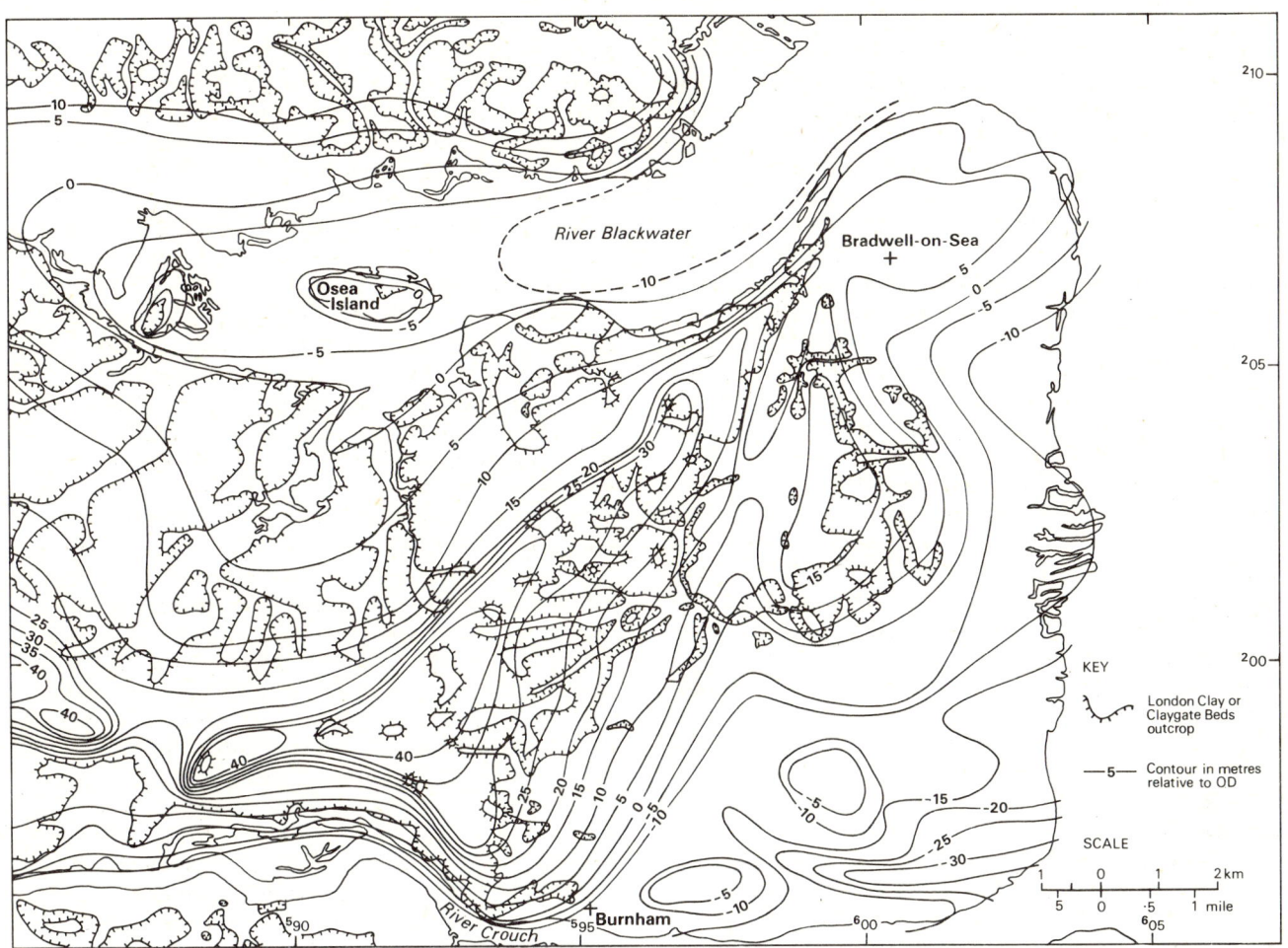

Fig. 3   The form of the surface of the bedrock (London Clay and Claygate Beds),
shown by contours plotted from 160 sample points

was being incised through the glacial sand and channel-fill silts and clays of the Dengie Peninsula, depositing fluvial gravels, which now occur as much-dissected remnants capping the older drift deposits or overlapping directly on to the bedrock.

Small patches of Second Terrace gravels with a basal bench at about 7 m above Ordnance Datum, are located mainly to the west of the marshes from Tillingham to Southminster, although the gravels underlying the southern part of the town of Burnham are composite deposits of the First and Second Terrace. Deposits of the First Terrace have a (basal) bench at approximately sea-level, and are present in the south around Burnham, extending for a short distance beneath the Marine or Estuarine Alluvium, and in the north-east, near Easthall Farm. The terrace deposits on Osea and Ramsey islands were probably laid down by a proto-Blackwater. Evidence from the three boreholes drilled in these deposits and from mapping suggests that they are more variable in lithology than the main spread of terrace material: they contain well-developed silt and clay seams particularly towards the west of Osea Island, although the interbedded gravels have a low fines content (see p. 9).

River Loam (Brickearth)
The River Loam or Brickearth is an orange, clayey, slightly sandy silt of uniform texture, commonly with roots and rare gravel occurring in stringers. This deposit is fairly restricted in areal extent and is confined mainly to the southern half of the region, where, in places, it overlies gravels of the Second and Third terraces, over-lapping on to the underlying bedrock. Elsewhere Brickearth occurs in isolated patches overlying gravels, but rarely attains a thickness of one metre, and has therefore not been shown on the geological map.

Grain-size distribution in some Brickearth samples, mainly from the area south of the Crouch, is typical of loess deposits. However, sedimentary features such as ripple lamination or feint horizontal banding, seen at some exposures, indicate subsequent deposition by water. These features are usually much disturbed by roots.

Marine or Estuarine Alluvium
Almost one quarter of the area of the Dengie Peninsula is covered by Marine or Estuarine Alluvium. The main expanse of this deposit forms the low-lying and extensive marshes flanking the sea coast, below which the London Clay bedrock falls away to the south-east to depths of more than 20 m below Ordnance Datum. Thin, small patches of Marine or Estuarine Alluvium occur along the south bank of the Blackwater, and along the north bank of the Crouch. The marshes were formed by salt-marsh aggradation in the past few centuries, and now have a hardened crust of varying thickness, usually of one or two metres, formed by desiccation.

Two types of Alluvium can be distinguished on the basis of lithology, the older being soft grey or blue-grey clays and silts with shells, organic material (which is sometimes so abundant as to colour the clays black), reed beds, and occasional peats or peaty clays. Cores from three Delft boreholes, which produce continuous undisturbed cores in a plastic sleeve, have shown bioturbation and laminations within the silts and clays. It is thought that these deposits were laid down in an upper tidal-flat, estuarine environment, the peat and reeds indicating salt-marsh conditions and the micro-fossils suggesting proximity of fresh-water inflow, producing reduced salinity. They crop out to the west of a fossil shell ridge, which crosses the marshes running north to south.

On the seaward side of this ridge, the clays and silts are overlain by soft pale brown to grey or blue-grey laminated silty sands and sandy silts - a younger Alluvium. The sedimentological and faunal characteristics of these deposits indicate a lower tidal-flat, estuarine environment: Cardium, whole and fragmented, is abundant throughout. Fossil shell ridges, similar to the migrating banks found on the tidal flats just off the coast at the present day, occur as clayey shell beds within the Alluvium, as shown in borehole TQ 99 NE 26. A line of the more recent banks occurs as a N-S trending ridge at the surface, separating the upper tidal-flat silts and clays from the more marine silts and sands.

The two types of Alluvium may interdigitate, or one may be absent, so that a simple coarsening-upwards sequence, indicative of one phase of marine transgression, is rarely seen. Firm, over-consolidated beds of silt and clay were encountered in some boreholes within the much softer sediments forming the bulk of the deposits. Such lithologies frequently have a brownish tinge, and were probably exposed at a time of temporary marine regression (Greensmith and Tucker, 1971, 1973).

The Marine or Estuarine Alluvium deposits occurring on the south bank of the Blackwater and on the north bank of the Crouch are lithologically similar to the clayey and silty alluvium of the marshes, differing only in their thickness. The deposits flanking the Blackwater vary between 1 m and 2 m in thickness, while those bordering the Crouch, a deeper channel than the Blackwater, rarely exceed 7 m to the west of Creeksea Place [935 962]. However, to the east, borehole TQ 99 NW 31 proved 12.7 m of Alluvium with sub-alluvial gravel, a thickness comparable with those proved in boreholes drilled on the marshes.

Marine Beach and Tidal-Flat Deposits (present day)
The present-day tidal-flats occur outside the area studied, mainly along the east coast and along the south bank of the Blackwater in the vicinity of Osea Island. These deposits are mainly laminated clay and silt with some sand and pockets of shell debris.

Storm Gravel Beach Deposits
Banks of gravel consisting mainly of flint with some shells and shell debris occur above the high-water mark on the exposed sections of the north bank of the Blackwater. They are parti-cularly extensive on Ramsey and Osea islands and in the vicinity of Stansgate Abbey Farm [931 058].

River Alluvium

Present-day river deposits of brown clayey silts are generally thin and of limited lateral extent. They occur in the valleys of some of the larger streams in the area.

The sub-Drift surface

The form of the bedrock (sub-Drift) surface of London Clay and Claygate Beds is shown by contours plotted at 5-m intervals in Fig. 3. In this area London Clay passes up into the Claygate Beds at about +30 m above Ordnance Datum. The prominent Tillingham-Althorne ridge is flanked to the south by a deep, well-defined buried channel beneath the present River Crouch, and to the north by a shallower drift-filled channel with remnant 'islands' of bedrock underlying the modern estuary of the Blackwater. In addition to these features, which influence the present scenery, boreholes have revealed other buried channels over $\frac{1}{2}$ km wide in the London Clay.

A SSW-NNE trending channel near Burnham is not obvious from the London Clay surface contours (Fig. 3), because their pattern has been modified by later channelling. This depression was identified by its distinctive channel-fill deposits (see p. 5), and confirmed by resistivity surveys. Boreholes have proved that the bedrock lies at nearly 5 m below Ordnance Datum to the north-east of Burnham, but resistivity work (see Appendix H) has shown that the deepest part of the channel may lie at a lower level.

A well-defined, erosional feature trending SSW-NNE in the Bradwell area, is infilled with similar thick distinctive channel-fill deposits near Tillingham, but becomes more difficult to trace to the north, near Orplands [998 063]. This feature has a narrow cross-section and steep channel walls (graded at about 1 in 4 in places: regarded as stable only in permafrost conditions). These are characteristics of a channel cut by glacial meltwater.

Several, probably fluvio-glacial, channels, running from west to east across the suballuvial bedrock, have also been identified from borehole information. The deep narrow feature beneath Great West Wick [985 967] and East Wick [002 965] (see Fig. 3), ending abruptly and steeply towards the west, also has the features of a channel cut in a glacial environment.

COMPOSITION OF THE SAND AND GRAVEL

The main constituent of all the gravels of the Dengie area is flint, subangular pebbles and rounded reworked pebbles occurring in approximately equal proportions. Cobble-size material is rare, although it has been encountered near the base of some of the deposits. Fine gravel, i.e. material in the 4 mm to 16 mm size range, generally makes up about two-thirds by weight of the gravel fraction.

Greensand sandstone and chert are the major accessory constituents, the amount present (5 to 20 per cent) varying somewhat with the type of deposit, as does the vein-quartz and quartzite content (5 to 15 per cent). The accessories are usually contained predominantly in the fine gravel-size fraction, the quartz and quartzite pebbles being subrounded to rounded, while the Greensand pebbles are usually subangular. The sand is mainly of medium grade throughout, consisting principally of clear quartz, although weathered and reworked flint chips predominate in the coarse fraction.

The proportions of gravel, sand and fines in the exposed terrace gravels and the suballuvial gravels are very similar: for example, the proportion of gravel in the potentially workable sand and gravel of the First, Second and Third terraces is 32 per cent while in the suballuvial gravels it is 29 per cent; the fines content of the terrace deposits is 11 per cent and of the sub-alluvial gravels 12 per cent. However, because the terrace deposits here include the 'clayey' pebbly sands of possible glacial origin, the true terrace deposits may contain an appreciably higher percentage of gravel and lower percentage of fines than the exposed mineral as a whole, although their composition has not been calculated separately.

The composition of the terrace gravels of Osea and Ramsey islands (see p. 8), sampled in boreholes TL 90 NW 3 and NW 5, differs from the remaining terrace gravels in containing considerably more gravel (62 per cent compared with 32 per cent) and less fines (3 per cent compared with 11 per cent). The composition of the mineral varies significantly from that of the other terrace deposits, in that quartz pebbles are more abundant in the deposits of the Blackwater area and Greensand pebbles are rarer. These differences suggest a probable derivation from the fluvio-glacial gravels to the north.

Fourth Terrace or kame gravels are more 'clayey' and less gravelly than the younger terrace deposits, the gravel content being 21 per cent and the fines 19 per cent. The coarse material is made up mainly of subangular and rounded flints with very little quartz, but with a higher proportion of Greensand than has been encountered in other terrace deposits of the area. However, their composition and altitude is comparable with that of the high-level gravels of unknown age in the Rayleigh area to the south-east (Gruhn, Bryan and Moss, 1974).

The specific gravity, water absorption and 10 per cent fines values of random samples of gravel from four sites were determined according to BS 812:1967 (see Table 2). The average water absorption value is well below 10 per cent, the maximum permissible value for dense aggregates in concrete making (Lea, 1970, p. 565), while according to British Standard 1201 (1965) a 10 per cent fines value of 15 tons is the minimum required for most uses of coarse aggregates, and 8 tons, the minimum value for gravel aggregates for surface dressings for roads (British Standard 1984, 1967).

Table 2. Results of 10 per cent fines, specific gravity and water absorption tests

Location of sample	Geological classification	Specific gravity			Absorption (per cent)	10 per cent fines value (tons)
		Apparent	Saturated surface dry	Oven dry		
959 993	Third Terrace	2.63	2.51	2.43	3.08	28
973 018	Third Terrace	2.62	2.55	2.51	1.63	23
9544 9832	Third Terrace	2.63	2.53	2.47	2.47	28
9424 0582	Second Terrace	2.63	2.54	2.48	2.33	27

THE MAP

The sand and gravel resource map is folded into the pocket at the end of this report. The base map is the Ordnance Survey 1:25 000 Outline Edition in grey, on which the topography is shown by contours in green, the geological data in black and the mineral resource information in shades of red.

Geological data
The geological boundary lines, symbols, etc., shown are taken from the geological map of this area, which was surveyed recently at the scale of 1:10 560. This information was obtained by detailed application of field mapping techniques by the field staff in the Institute's East Anglia and South-East England Unit. Borehole data, which include the stratigraphic relations and mean particle-size distribution of the sand and gravel samples collected during the assessment survey, are also shown.

The geological boundaries show the best available interpretation of the information available at the time of survey. However, it is inevitable that local irregularities or discrepancies will be revealed by some boreholes (for example, at boreholes TQ 99 NW 27 and TQ 99 NE 18). These are taken into account in the assessment of resources (see below and Appendix B).

Mineral Resource Information
The mineral-bearing ground is sub-divided into resource blocks (see Appendix A). Within a resource block the mineral is sub-divided into areas where it is 'exposed' and areas where it is present in continuous (or almost continuous) spreads beneath overburden. The mineral is identified as 'exposed' where the overburden, commonly consisting only of soil and sub-soil, averages less than 1.0 m (3.5 ft) in thickness.

Beneath overburden the mineral may be continuous (or almost continuous) or discontinuous. The recognition of these categories is dependent upon the importance attached to the proportion of boreholes which did not find potentially workable sand and gravel and the distribution of barren boreholes within a block. The mineral is described as 'almost continuous' if it is present in

75 per cent or more of the boreholes in a resource block. The 'discontinuous' category has also been recognised on the present sheet.

Areas where bedrock crops out, where boreholes indicate absence of sand and gravel beneath cover and where sand and gravel beneath cover is interpreted to be not potentially workable are uncoloured on the Map; where appropriate the relevant criterion is noted. In such areas it has been assumed that mineral is absent except in infrequent and relatively minor patches which can neither be outlined nor assessed quantitatively in the context of this survey. Areas of unassessed sand and gravel, for example, areas of storm gravel beach deposits, are indicated by a red stipple.

The area of the exposed sand and gravel is measured from the mapped geological boundary lines. The whole of this area is considered as mineral, although it may include small areas where sand and gravel is not present or is not potentially workable. Inferred boundaries have been inserted where sand and gravel is interpreted to be not potentially workable or absent. Such boundaries (for which a distinctive symbol is used) are drawn primarily for the purpose of volume estimation. The symbol is intended to convey an approximate location within a likely zone of occurrence rather than to represent the breadth of the zone, its size being limited only by cartographic considerations. For the purpose of measuring areas the centre-line of the symbol is used.

RESULTS

The statistical results are summarised in Table 3. Fuller grading particulars are shown in Fig. 5.

Accuracy of Results
For the three resource blocks the accuracy of the results at the symmetrical 95 per cent probability level varies between 33 per cent and 45 per cent (that is, it is probable that nineteen times out of twenty the true volumes present lie within these limits). However, the true values are more likely to be nearer the figures estimated than the limits. Moreover, it is probable that in each

Table 3. The sand and gravel resources of the Dengie Peninsula

	Area		Mean thickness				Volume of mineral				Mean grading percentage		
	Block	Mineral	Overburden		Mineral				Limits at 95% confidence level		Fines	Sand	Gravel
	km²	km²	m	ft	m	ft	Million m³	Million yd³	±%	± Volume million m³	-1/16 mm	+1/16 mm -4 mm	+4 mm
A	29.3	12.3	1.2	3.9	3.0	9.8	36.4	47.7	33	12.0	9	58	33
B	111.3	13.2	1.3	4.3	2.8	9.2	37.4	49.0	44	16.5	12	52	36
C	41.8	13.9	8.9	29.1	4.6	15.0	64.1	84.0	45	28.8	12	59	29
A + B	140.6	25.5	1.2	3.9	2.9	9.5	73.7	96.5	25	18.4	10	55	35
Total	182.4	39.4	3.0	9.8	3.3	10.8	137.9	180.6	22	30.3	11	55	34

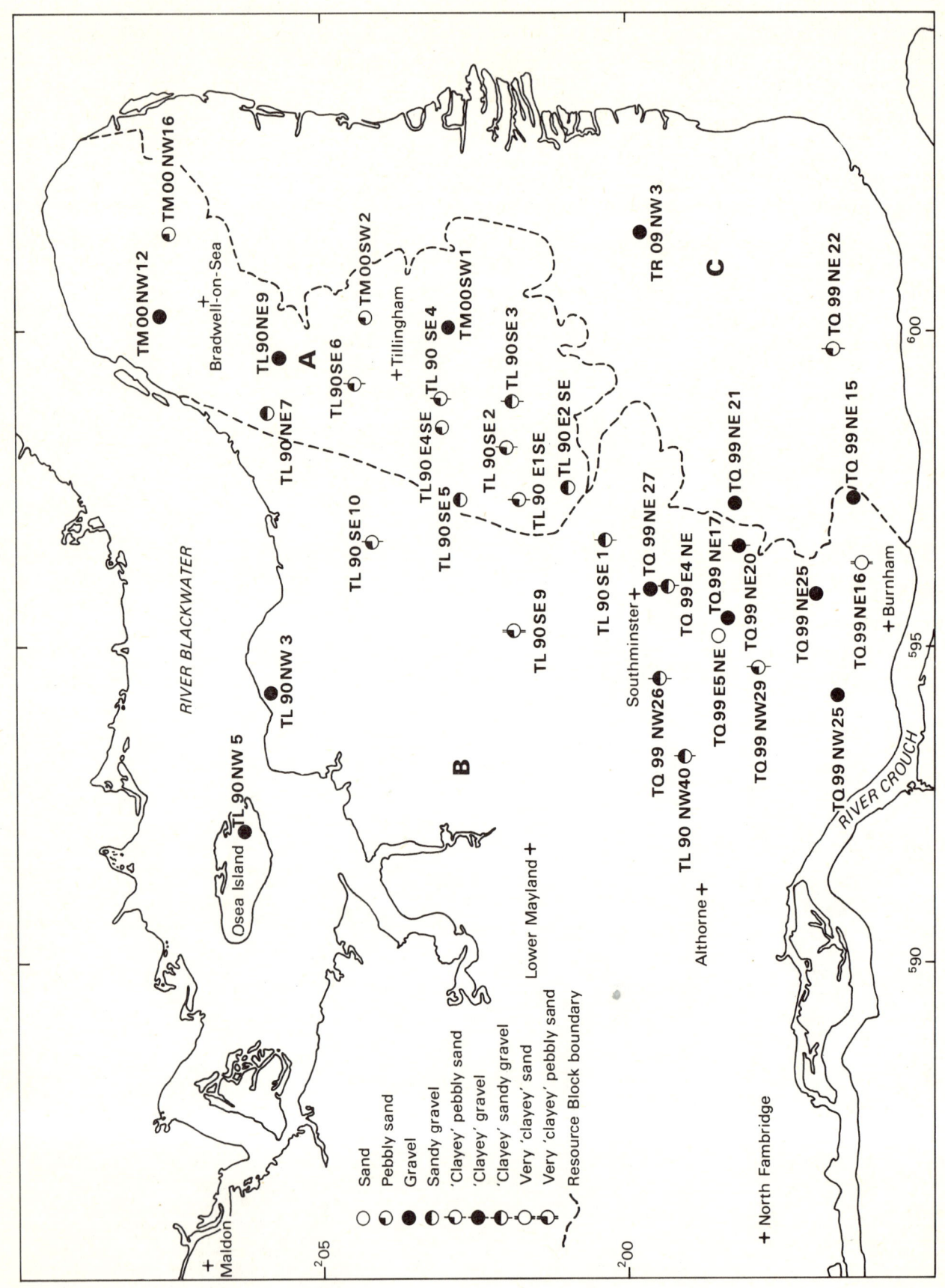

Fig. 4 Regional grading characteristics of the mineral based on 29 mineral assessment boreholes and five exposures

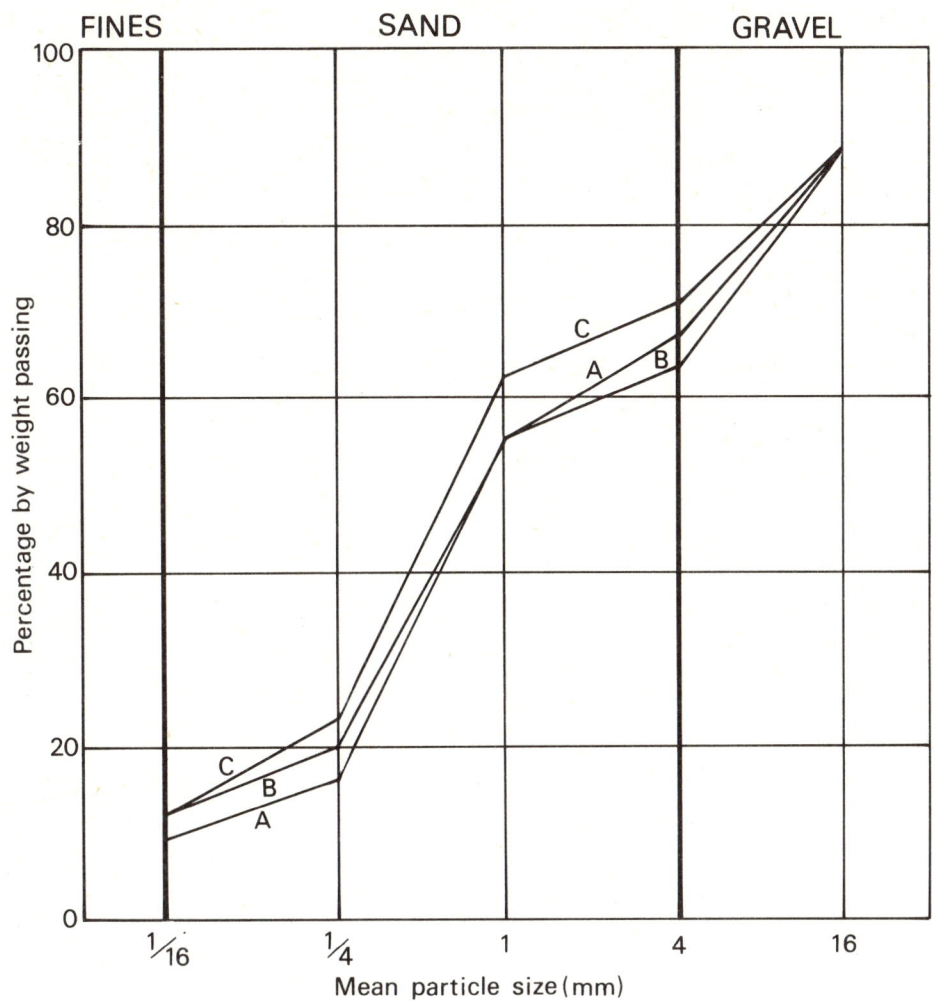

Block	Percentage by weight passing				
	1/16 mm	$\frac{1}{4}$ mm	1 mm	4 mm	16 mm
A	9	16	56	67	88
B	12	20	56	64	87
C	12	23	63	71	89

Fig. 5   Particle-size distribution for the assessed thickness of mineral in resource blocks A, B and C

block roughly the same percentage limits would apply for the estimate of volume of a very much smaller parcel of ground (say, 200 acres) containing similar sand and gravel deposits if the results from the same number of sample points (as provided by, say, ten boreholes) were used in the calculation. Thus, if closer limits are needed for the quotation of reserves of part of a block, it can be expected that data from more than ten sample points will be required, even if the area is quite small. This point can be illustrated by considering the whole of the potentially workable sand and gravel on this sheet. The volume (137.9 million m^3) can be estimated to limits of $\pm$22 per cent at the 95 per cent probability level, by a calculation based on the data from 43 sample points spread across the three resource blocks.

However, it must be emphasised that the quoted volume of sand and gravel has no simple relationship with the amount that could be extracted in practice, as no allowance has been made in the calculations for any restraints (such as existing buildings and roads) on the use of the land for mineral workings.

NOTES ON RESOURCE BLOCKS A TO C

Block A

Terrace deposits, classified as mineral, cover more than one-third of the area of block A in rather thin, discontinuous patches overlying London Clay, or occasionally Head derived mainly from London Clay. London Clay, obscured in part by Head, crops out over the remainder of the block, and is overlain by thin River Alluvium in some minor stream valleys and by Marine or Estuarine Alluvium near the northern coastline.

The gravels with one exception are of the Second and Third Terraces. (In these notes, the terms "Second Terrace" and "Third Terrace" also include material of glacial meltwater origin, as the two types of deposits cannot, in general, be distinguished by surface mapping). The exception, a patch of First Terrace [017 080] in the extreme north-east of the peninsula, is the sole representative of this terrace. It has not been sampled. The Second Terrace gravels are confined to the area around Bradwell and to the east of Tillingham, while the bulk of the Third Terrace deposits are in the southern part of the block, covering most of the ground around Asheldham, Dengie and Tillingham, where they continue to be fairly extensively worked.

The only mapped occurrence of River Terrace Loam near Small Gains [0070 0212] is thought to overlie London Clay except at its southern limit, where Second Terrace gravels extend for a short distance beneath it. In some small valleys, for example, to the west of Reddings [981 030] and to the south of Shingleford [004 043], the terrace gravels are overlain by Head.

Hand augering demonstrated the continuity of the gravel in the valley between Asheldham and Dengie village with the fairly thick deposits of gravel near the edges of the terrace outcrop on either side of the valley, proved in borehole

TL 90 SE 2. An uncompleted resistivity traverse across the valley running south from Bradwell Waterside [996 079] had indicated the probable existence of gravels beneath the outcropping Head.

Mineral proved in borehole TL 90 NE 7 is thought to be the basal lag gravel deposit of a subglacial stream channel. The limits of such a deposit are speculative. However, a thin layer of silty gravelly sand lying between the Head and the underlying channel deposits may be the westward extension of the Third Terrace deposits exposed in the vicinity of Curry [9968 0570]. Mineral, as either channel or terrace deposits, is shown on the map to be continuous between there and borehole TL 90 NE 7. Channel-lag gravels exist farther south in the buried channel, as proved by borehole TL 90 SE 7, but at too great a depth to be considered as mineral.

All of the 19 assessment boreholes drilled in block A proved bedrock, but only 11 encountered mineral; its thickness ranged from 1.1 m in borehole TM 00 SW 2 to 5.8 m in TL 90 NE 9. The mean thickness of mineral, taking into account other boreholes, is 3.0 m. Hand augering has shown lateral variations in the grade of the deposit to be local.

Overburden, mostly top-soil, was generally found to be less than 1 m thick; however, owing to the high value obtained from TL 90 NE 7, where 9.0 m of Head and channel-fill deposits overlie mineral, the mean thickness is 1.2 m.

The mean grading for the block is fines 9 per cent, sand 58 per cent, gravel 33 per cent, and the total volume of mineral was calculated to be 36.4$\pm$12.0 million m^3 (at the 95 per cent probability level).

Block B

Mineral is found mainly in the south-eastern part of this block, a large area being barren. The total area of block B is just over 111 km^2, of which sand and gravel classified as mineral covers 13.2 km^2. London Clay is the bedrock over most of the block and crops out nearly everywhere to the east of the Tillingham - Althorne ridge, except where covered by London Clay Head and Marine or Estuarine Alluvium near the river banks. Claygate Beds crop out at the top of the southern part of the ridge on the north bank of the Crouch, except where they are concealed by Fourth Terrace or kame gravels near Mayland and Althorne. Terrace deposits, probably of a proto-Blackwater, overlie London Clay on Ramsey and Osea islands, the coasts of which are flanked by storm-beach deposits which have not been sampled in the present survey and are not included in the assessment of resources.

As with block A, most of the mineral is Third Terrace gravel, which covers most of the ground in the area around Southminster and to the west and north of Burnham-on-Crouch. The gravels are being worked in several pits to the south of Southminster (Appendix J). The only established occurrence of Second Terrace deposits is to the north-east of Southminster: the gravels of Burnham are mapped as undifferentiated First and Second Terrace. Neither the Second Terrace nor the composite First/Second Terrace has been sampled. Gravels mapped as First Terrace deposits form the small features flanking the marshes to the east and north-east of Burnham. The gravels encountered in borehole TQ 99 NE 20 have been classified as Head, which was probably derived from the Third Terrace to the west. First Terrace Sand and Gravel overlie channel deposits in places, for example, to the west of Dammer Wick [962 969], in the areas of Newman's Farm [962 974], Burnham Wick [962 958] and the southern part of Burnham. To the east of Burnham, Head Brickearth overlies mineral, which may be terrace or channel deposits, and the First Terrace gravels extend to the east beneath the Head and Marine or Estuarine Alluvium, as shown by borehole TQ 99 NE 15.

River Loam overlies the Third Terrace gravels in several places, as proved by boreholes TQ 99 NW 25 and 258/9b, and the relationship of the overburden to the underlying mineral can be seen at exposures [9512 9851] at the pits at Ratsborough.

Thirty boreholes were drilled in block B in this survey, of which 15 proved mineral and two TQ 99 NE 20 and TQ 99 NE 24 did not reach bedrock. Mineral thickness varied from 1.3 m in boreholes TQ 99 NW 40, TL 90 SE 10 and TL 90 SE 9, to 10.0 m in borehole TL 90 SE 1, the mean value, calculated from all available borehole data being 2.8 m. Boreholes TQ 89 NE 10, TQ 99 NW 27, TQ 99 NE 18 and TL 90 NW 4 proved that the drift deposits were non-mineral.

The average thickness of the overburden was 1.3 m. Five boreholes proved topsoil or made ground overlying the gravels, and of the remainder, one borehole, TQ 99 NE 25 passed through a total of 5.7 m of made ground, Head Brickearth and fine-grained channel-fill deposits overlying mineral.

The mean grading for mineral in block B is fines 12 per cent, sand 52 per cent and gravel 36 per cent, the total volume of mineral being 37.4±16.5 million m^3.

Claygate Beds were sampled in boreholes TQ 89 NE 10, TQ 99 NW 27 and TL 90 SW 1, and were found to contain only thin laminae of fine sand in clay.

Several boreholes were sunk in the Marine or Estuarine Alluvium on the banks of the Crouch and Blackwater, but of those that proved sand and gravel of mineral grade, none showed a thickness of more than about 0.8 m.

Block C

The western limits of block C are defined by the edge of the Marine or Estuarine Alluvium, which forms the coastal marshes. The mineral in this block is confined to the south, and is overlain everywhere by Alluvium. As the extent of the mineral beneath the overburden is uncertain, only the category 'discontinuous mineral beneath overburden' has been outlined on the resource map to include those boreholes which proved mineral. Areas of discontinuous mineral cover one-third (13.9 km^2) of the total block area of almost 42 km^2.

Of the 18 assessment boreholes drilled in block C, one, TR 09 NW 6 did not reach bedrock and only 3, TQ 99 NE 21, TQ 99 NE 22 and TR 09 NW 3 proved mineral (although others proved some sand and gravel). Mineral thicknesses are very variable, ranging from 2.3 m in the Ministry of Public Buildings and Works borehole TM 00 SW 8 to more than 12.2 m in Hydrogeological Department record 259/8b. The mean thickness is calculated to be 4.6 m with a mean overburden thickness of 8.9 m.

Mineral was thought to be present at the sites of borehole TR 09 NW 6 and Dutch probes TR 09 NW 13, TQ 99 NE 30 and TR 09 NW 15, although the drilling was stopped by technical problems before bedrock was reached. The results of these holes were not used in the calculation of resources.

The mean grading for block C is fines 12 per cent, sand 59 per cent, gravel 29 per cent and the total volume is 64.1±28.8 million m^3. The fine, blue-grey alluvium, frequently encountered in this area, was sampled at several places but was found to contain too high a proportion of fines to be classified as mineral.

Although the suballuvial gravels are present in quality and quantities comparable with the exposed terrace gravels, their depth of burial and the uncertainty of the distribution of the deposits makes them unlikely to be economically workable.

APPENDIX A: FIELD AND LABORATORY PROCEDURES

Trial and error during initial studies of the complex and variable glacial deposits of East Anglia and Essex showed that an absolute minimum of five sample points evenly distributed across the sand and gravel are needed to provide a worthwhile statistical assessment, but that, where possible, there should be not less than ten. Sample points are any points for which adequate information exists about the nature and thickness of the deposit and may include boreholes other than those drilled during the survey, Dutch Probes and exposures. In particular, the cooperation of sand and gravel operators ensures that the boreholes are not drilled where reliable information is already available; although this may be used in the calculations, it is held confidentially by the Institute and cannot be disclosed.

The mineral shown on each 1:25 000 sheet is divided into resource blocks. The arbitrary size selected, approximately 10 km², is a compromise to meet the aims of the survey by providing sufficient sample points in each block. As far as possible the block boundaries are determined by geological boundaries so that, for example, suballuvial and other river gravels are separated. Otherwise division is by arbitrary lines, which may bear no relationship to the geology. The blocks are drawn provisionally before drilling begins.

A reconnaissance of the ground is carried out to record any exposures and inquiries are made to ascertain what borehole information is available. Borehole sites are then selected to provide an even pattern of sample points at a density of approximately one per square kilometre. However, because broad trends are independently overlain by smaller scale characteristically random variations, it is unnecessary to adhere to a square grid pattern. Thus such factors as ease of access and the need to minimise disturbance to land and the public are taken into account in siting the holes; at the same time it is necessary to guard against the possibility that ease of access (that is, the positions of roads and farms) may reflect particular geological conditions, which may bias the drilling results.

The drilling machine employed should be capable of providing a continuous sample representative of all unconsolidated deposits, so that the in-situ grading can be determined, if necessary, to a depth of 30 m (100 ft) at a diameter of about 200 mm (8 in), beneath different types of overburden. It should be reliable, quiet, mobile and relatively small (so that it can be moved to sites of difficult access). Shell and auger rigs have proved to be almost ideal.

The rigs are modified to enable deposits above the water table to be drilled 'dry', instead of with water added to facilitate the drilling, to minimise the amount of material drawn in from outside the limits of the hole. The samples thus obtained are representative of the in-situ grading, and satisfy one of the most important aims of the survey. Below the water table the rigs are used conventionally, although this may result in the loss of some of the fines fraction and the pumping action of the bailer tends to draw unwanted material into the hole from the sides or the bottom.

A continuous series of bulk samples is taken throughout the sand and gravel. Ideally samples are composed exclusively of the whole of the material encountered in the borehole between stated depths. However, care is taken to discard, as far as possible, material which has caved or has been pumped from the bottom of the hole. A new sample is commenced whenever there is an appreciable lithological change within the sand and gravel, or at every 1 m (3.3 ft) depth. The samples, each weighing between 25 and 45 kg (55 and 100 lb), are despatched in heavy duty polythene bags to a laboratory for grading. The grading procedure is based on British Standard 1377 (1967). Random checks on the accuracy of the grading are made in the laboratories of the Institute's Geochemical Division.

All data, including mean grading analysis figures calculated for the total thickness of the mineral, are entered on standard record sheets, abbreviated copies of which are reproduced in Appendix G.

A resistivity survey using an A.B.E.M. Terrameter in its A.C. version, was included in this assessment programme. Details are given in Appendix H.

Detailed records may be consulted at the appropriate offices of the Institute, upon application to the Head, Mineral Assessment Unit.

APPENDIX B: STATISTICAL PROCEDURE

Statistical Assessment

1. A statistical assessment is made of an area of mineral greater than 2 km², if there is a minimum of five evenly spaced boreholes in the resource block (for smaller areas see paragraph 12 below).

2. The simple methods used in the calculations are consistent with the amount of data provided by the survey. Conventional symmetrical confidence limits are calculated for the 95 per cent probability level, that is, there is a 5 per cent or one in twenty chance of a result falling outside the stated limits.

3. The volume estimate (V) for the mineral in a given block is the product of the two variables, the sampled areas (A) and the mean thickness ($\bar{l}_m$) calculated from the individual thicknesses at the sample points. The standard deviations for these variables are related such that

$$S_V = \sqrt{(S_A{}^2 + S_{\bar{l}_m}{}^2)} \qquad [1]$$

4. The above relationship may be transposed such that

$$S_V = S_{\bar{l}_m} \sqrt{(1 + S_A{}^2/S_{\bar{l}_m}{}^2)} \qquad [2]$$

From this it can be seen that as $S_A{}^2/S_{\bar{l}_m}{}^2$ tends to 0, S_V tends to $S_{\bar{l}_m}$.
If, therefore, the standard deviation for area is small with respect to that for mean thickness, the standard deviation for volume approximates to that for mean thickness.

5. Given that the number of approximately evenly spaced sample points in the sampled area is n with

mineral thickness measurements $l_{m_1}, l_{m_2}, \ldots l_{m_n}$, then the best estimate of mean thickness, $\bar{l}_m$, is given by

$$\frac{\sum (l_{\bar{m}_1} + l_{m_2} \ldots l_{m_n})}{n}$$

For groups of closely spaced boreholes a discretionary weighting factor may be applied to avoid bias (see note on weighting below). The standard deviation for mean thickness, $S_{\bar{l}}$, expressed as a proportion of the mean thickness is given by

$$S_{\bar{l}} = (1/\bar{l}_m) \sqrt{[(l_m - \bar{l}_m)^2/(n-1)]}$$

where l_m is any value in the series l_{m_1} to l_{m_n}.

6. The sampled area in each resource block is coloured pink on the map. Wherever possible, calculations relate to the mineral within mapped geological boundaries (which may not necessarily correspond to the limits of deposit). Where the area is not defined by a mapped boundary, that is, where the boundary is inferred, a distinctive symbol is used. Experience suggests that the errors in determining area are small relative to those in thickness. The relationship $S_A/S_{\bar{l}_m} \leqslant \frac{1}{3}$ is assumed in all cases. It follows from equation [2] that

$$S_{\bar{l}_m} \leqslant S_V \leqslant 1.05 \, S_{\bar{l}_m} \qquad [3]$$

7. The limits on the estimate of mean thickness of mineral, $L_{\bar{l}_m}$, may be expressed in absolute units $\pm(t/\sqrt{n}) \times S_{\bar{l}_m}$ or as a percentage $\pm(t/\sqrt{n}) \times S_{\bar{l}_m} \times (100/\bar{l}_m)$ per cent, where t is Student's t at the 95 per cent probability level for $(n-1)$ degrees of freedom, evaluated by reference to statistical tables. (In applying Student's t it is assumed that the measurements are distributed normally).

8. Values of t at the 95 per cent probability level for values of n up to 20 are as follows:

n	t	n	t
1	infinity	11	2.228
2	12.706	12	2.201
3	4.303	13	2.179
4	3.182	14	2.160
5	2.776	15	2.145
6	2.571	16	2.131
7	2.447	17	2.120
8	2.365	18	2.110
9	2.306	19	2.101
10	2.262	20	2.093

(from Table 12, Biometrika Tables for Statisticians, Volume 1, Second Edition, Cambridge University Press, 1962). When n is greater than 20, 1.96 is used (the value of t when n is infinity).

9. In calculating confidence limits for volume, L_V, the following inequality corresponding to equation [3] is applied: $L_{\bar{l}_m} \leqslant L_V \leqslant 1.05 L_{\bar{l}_m}$

10. In summary, for values of n between 5 and 20, L_V is calculated as

$$[(1.05 \times t)/\bar{l}_m] \times [\sqrt{\Sigma(l_m - \bar{l}_m)^2/n(n-1)}] \times 100$$
per cent, and when n is greater than 20, as

$$[(1.05 \times 1.96)/\bar{l}_m] \times [\sqrt{\Sigma(l_m - \bar{l}_m)^2/n(n-1)}] \times 100$$
per cent.

11. The application of this procedure to a fictitious area is illustrated in Figs. 6 and 7.

12. If the area of mineral in a resource block is between 0.25 km² and 2 km² an assessment is inferred, based on geological and topographical information usually supported by the data from one or two boreholes. The volume of mineral is calculated as the product of the area, measured from field data, and the estimated thickness. Confidence limits are not calculated.

13. In some cases a resource block may include an area left uncoloured on the map, within which mineral (as defined) is interpreted to be generally absent. If there is reason to believe that some mineral may be present, an inferred assessment may be made.

14. No assessment is attempted for an isolated area of mineral less than 0.25 km².

15. *Note on Weighting* The thickness of a deposit at any point may be governed solely by the position of the point in relation to a broad trend. However, most sand and gravel deposits also exhibit a random pattern of local, and sometimes considerable, variation in thickness. Thus the distribution of sample points need be only approximately regular and in estimating the mean thickness only simple weighting is necessary. In practice, equal weighting can often be applied to thicknesses at all sample points. If, however, there is a distinctly unequal distribution of points, bias is avoided by dividing the sampled area into broad zones, to each of which a value roughly proportional to its area is assigned. This value is then shared between the data points within the zone as the weighting factor.

APPENDIX C: CLASSIFICATION AND DESCRIPTION OF SAND AND GRAVEL

For the purposes of assessing resources of sand and gravel a classification should take account of economically important characteristics of the deposit, in particular the absolute content of fines and the ratio of sand to gravel.

The terminology commonly used by geologists when describing sedimentary rocks (Wentworth, 1922) is not entirely satisfactory for this purpose. For example, Wentworth proposed that a deposit should be described as a 'gravelly sand' when it contains more sand than gravel and there is at least 10 per cent of gravel, provided that there is less than 10 per cent of material finer than sand (less than $\frac{1}{16}$ mm) and coarser than pebbles (more than 64 mm in diameter). Because deposits containing more than 10 per cent fines are not embraced by this system a modified binary classification based on Willman (1942) has been adopted.

When the fines content exceeds 40 per cent the material is not considered to be potentially workable and falls outside the definition of mineral. Deposits which contain 40 per cent fines or less are classified primarily on the ratio of sand to gravel but qualified in the light of the fines content, as follows: less than 10 per cent fines – no qualification; 10 per cent or more but less than 20 per cent fines – 'clayey'; 20 to 40 per cent fines – 'very clayey'.

The term 'clay' (as written, with single quote marks) is used to describe all material passing $\frac{1}{16}$ mm. Thus it has no mineralogical significance and includes particles falling within the size range of silt. The normal meaning applies to the term clay where it does not appear in single quotation marks.

The ratio of sand to gravel defines the boundaries between sand, pebbly sand, sandy gravel and gravel (at 19:1, 3:1 and 1:1).

Thus it is possible to classify the mineral into one of twelve descriptive categories (see Fig.8). The procedure is as follows:

1. Classify according to ratio of sand to gravel.
2. Describe fines.

For example, a deposit grading 11 per cent gravel, 70 per cent sand and 19 per cent fines is classified as 'clayey' pebbly sand. This short description is included in the borehole log (see Note 11, p. 22).

Many differing proposals exist for the classification of the grain size of sediments (Atterberg, 1905; Udden, 1914; Wentworth, 1922; Wentworth, 1935; Allen, 1936; Twenhofel, 1937; Lane and others, 1947). As Archer (1970a, b) has emphasised, there is a pressing need for a simple metric scale acceptable to both scientific and engineering interests, for which the class limit sizes correspond closely with certain marked changes in the natural properties of mineral particles. For example, there is an important change in the degree of cohesion between particles at about the $\frac{1}{16}$-mm size, which approximates to the generally accepted boundary between silt and sand. These and other requirements are met by a system based on Udden's geometric scale and a simplified form of Wentworth's terminology (Table 4), which is used in this Report.

The fairly wide intervals in the scale are consistent with the general level of accuracy of the qualitative assessments of the resource blocks. Three sizes of sand are recognised, fine ($-\frac{1}{4}+\frac{1}{16}$ mm), medium ($-1+\frac{1}{4}$ mm) and coarse ($-4+1$ mm). The boundary at 16 mm distinguishes a range of finer gravel ($-16+4$ mm), often characterised by abundance of worn tough pebbles of vein quartz, from larger pebbles often of notably different materials. The boundary at 64 mm distinguishes pebbles from cobbles. The term 'gravel' is used loosely to denote both pebble-sized and cobble-sized material.

The size distribution of borehole samples is determined by sieve analysis, which is presented by the laboratory as logarithmic cumulative curves (see, for example, British Standard 1377: 1967). In this report the grading is tabulated on the borehole record sheets (Appendix G), the intercepts corresponding with the simple geometric scale $\frac{1}{16}$ mm, $\frac{1}{4}$ mm, 1 mm, 4 mm, 16 mm and so on as required. Original sample grading curves are available for reference at the appropriate office of the Institute.

Each bulk sample is described, subjectively, by a geologist at the borehole site. Being based on visual examination, the description of the grading is inexact, the accuracy depending on the experience of the observer. The descriptions recorded are modified, as necessary, when the laboratory results become available.

The relative proportions of the rock types present in the gravel fraction are indicated by the use of the words 'and' or 'with'. For example, 'flint and quartz' indicates very approximate equal proportions with neither constituent accounting for less than about 25 per cent of the whole; 'flint with quartz' indicates that flint is dominant and quartz, the principal accessory rock types, comprises 5 to 25 per cent of the whole. Where the accessory material accounts for less than 5 per cent of the whole, but is still readily apparent, the phrase 'with some' has been used. Rare constituents are referred to as 'trace'.

The terms used in the field to describe the degree of rounding of particles, which is concerned with the sharpness of the edges and corners of a clastic fragment and not the shape (after Pettijohn, 1957), are as follows.

Angular: showing little or no evidence of wear; sharp edges and corners.

Subangular: showing definite effects of wear. Fragments still have their original form but edges and corners begin to be rounded off.

Subrounded: showing considerable wear. The edges and corners are rounded off to smooth curves. Original grain shape is still distinct.

Rounded: original faces almost completely destroyed, but some comparatively flat surfaces may still remain. All original edges and corners have been smoothed off to rather broad curves. Original shape is still apparent.

Well-rounded: no original faces, edges or corners left. The entire surface consists of broad curves; flat areas are absent. The original shape is suggested by the present form of the grain.

Table 4   Classification of gravel, sand and fines

Size limits	Grain size description	Qualification	Primary classification
64 mm	Cobble		
16 mm	Pebble	Coarse	Gravel
4 mm		Fine	
1 mm	Sand	Coarse	Sand
$\frac{1}{4}$ mm		Medium	
$\frac{1}{16}$ mm		Fine	
	Fines (silt and clay)		Fines

Block Calculation | 1:25 000 } Fictitious
Block

Area
  Block: 11.08 km^2
  Mineral: 8.32 km^2

Volume
  Overburden: 21 million m^3
  Mineral: 54 million m^3

Mean Thickness
  Overburden: 2.5 m
  Mineral: 6.5 m

Confidence limits of the estimate of mineral volume
  at the 95 per cent probability level: ± 20 per cent
That is, the volume of mineral (with 95 per cent
  probability): 54 ± 11 million m^3

Thickness estimate: measurements in metres
l_o = overburden thickness l_m = mineral thickness

Sample point	Weighting w	Overburden l_o	wl_o	Mineral l_m	wl_m	Remarks
SE 14	1	1.5	1.5	9.4	9.4	IMAU boreholes
SE 18	1	3.3	3.3	5.8	5.8	
SE 20	1	nil	–	6.9	6.9	
SE 22	1	0.7	0.7	6.4	6.4	
SE 23	1	6.2	6.2	4.1	4.1	
SE 24	1	4.3	4.3	6.4	6.4	
SE 17	$\frac{1}{2}$	1.2 }	1.6	9.8 }	7.2	Hydrogeological Dept record
123/45	$\frac{1}{2}$	2.0 }		4.6 }		
1	$\frac{1}{4}$	2.7 }		7.3 }		Close group of four boreholes (commercial)
2	$\frac{1}{4}$	4.5 }	2.6	3.2 }	5.8	
3	$\frac{1}{4}$	0.4 }		6.8 }		
4	$\frac{1}{4}$	2.8 }		5.9 }		
Totals **Means**	$\Sigma w = 8$	$\Sigma wl_o = 20.2$ $\bar{l}_o = 2.5$		$\Sigma wl_m = 52.0$ $\bar{l}_m = 6.5$		

Calculation of confidence limits

l_m	$(l_m - \bar{l}_m)$	$(l_m - \bar{l}_m)^2$
9.4	2.9	8.41
5.8	0.7	0.49
6.9	0.4	0.16
6.4	0.1	0.01
4.1	2.4	5.76
6.4	0.1	0.01
7.2	0.7	0.49
5.8	0.7	0.49

$\Sigma(l_m - \bar{l}_m)^2 = 15.82$

$n = 8$

$t = 2.365$

L_V is calculated as

$$1.05 \times \frac{t}{\bar{l}_m} \sqrt{\frac{\Sigma(l_m - \bar{l}_m)^2}{n(n-1)}} \times 100$$

$$= 1.05 \times \frac{2.365}{6.5} \sqrt{\frac{15.82}{8 \times 7}} \times 100$$

$$= 20.3$$

$$\simeq 20 \text{ per cent}$$

Fig. 6   Example of resource block assessment: calculation and results

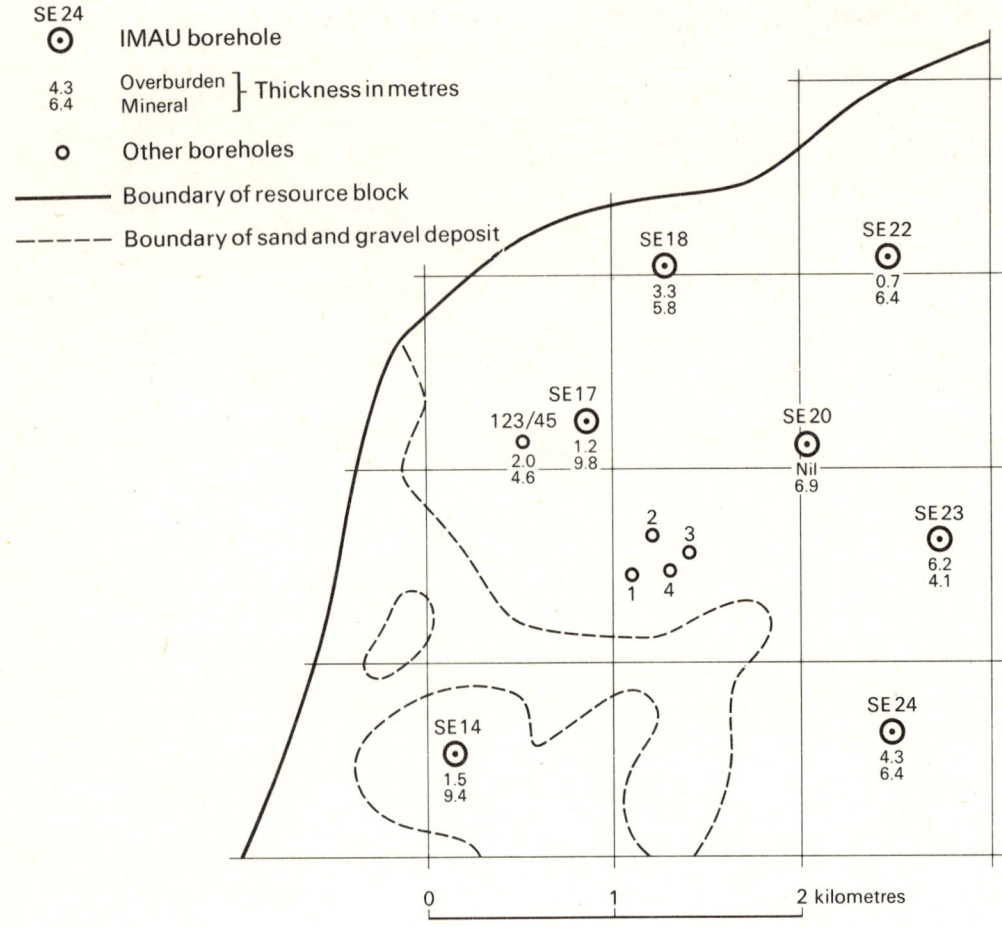

Fig. 7   Example of resource block assessment: map of fictitious block

I   Gravel

II   'Clayey' gravel

III   'Very clayey' gravel

IV   Sandy gravel

V   'Clayey' sandy gravel

VI   'Very clayey' sandy gravel

VII   Pebbly sand

VIII   'Clayey' pebbly sand

IX   'Very clayey' pebbly sand

X   Sand

IX   'Clayey' sand

IIX   'Very clayey' sand

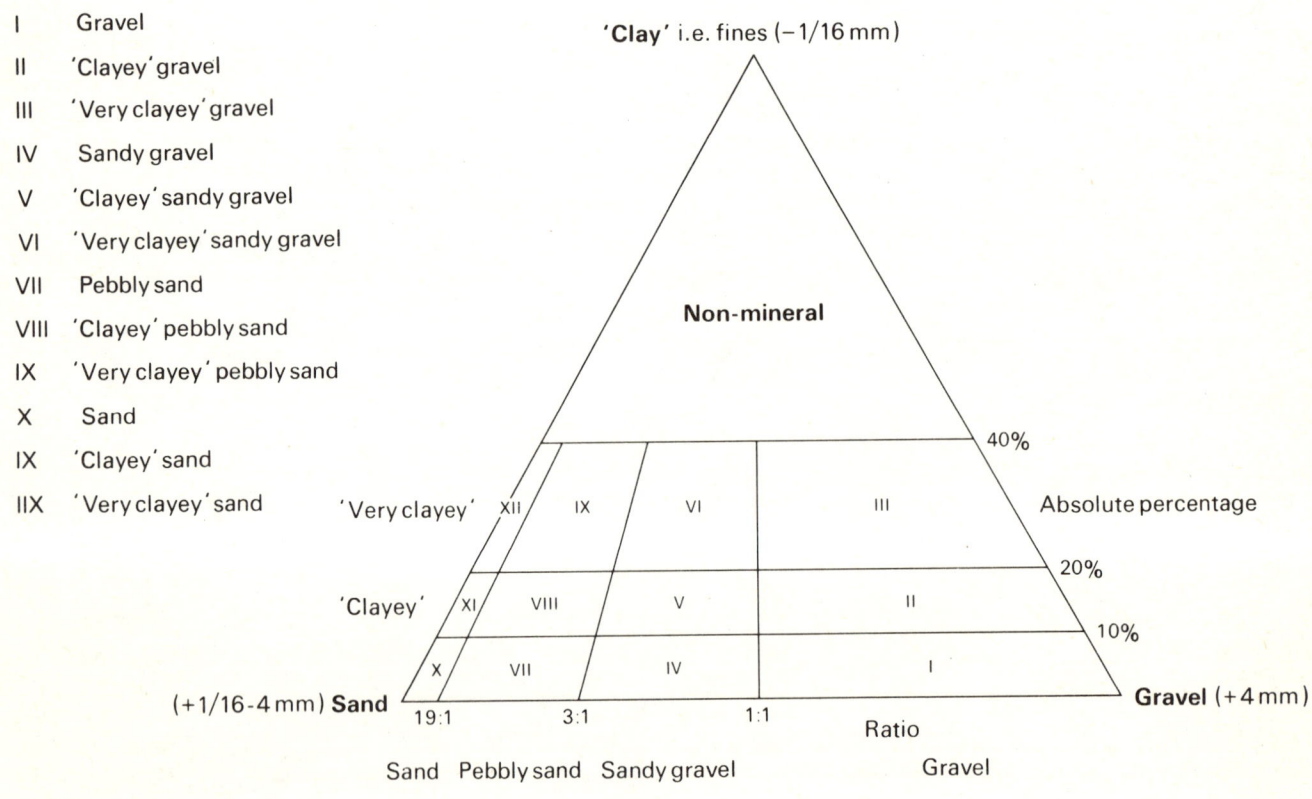

Fig. 8   Diagram to show the descriptive categories used in the classification of sand and gravel

20

APPENDIX D: EXPLANATION OF BOREHOLE RECORDS

ANNOTATED EXAMPLE

TR 09 NW 3[1]   0155 9974[2]   Dengie, Essex[3]   Block C

Surface level (+2.1 m) +7.0 ft[4]   [7]Overburden 4.2 m
Water struck at -2.1 m[5]   Mineral 2.6 m
Shell 203 mm diameter[6]   Waste 5.7 m
February 1973   Bedrock 0.3 m+[9]

LOG

Geological Classification[10]	Lithology[11]	Thickness m	Depth[8] m
Soil	Clayey silt with broken shells	0.2	0.2
Estuarine Alluvium	Sandy clay and silt with occasional shells and organic debris. Mainly medium bluish grey[12]	4.0	4.2
Buried Channel Deposits	Gravel Gravel becoming coarser with depth, being mainly fine grade near top of deposit, but having almost equal proportions of fine and coarse gravel	2.6	6.8
	Silty clay and clayey silt with occasional reeds and shell fragments. Soft to stiff, dark bluish grey with some olive grey and yellowish brown bands from 10.2 m to 11.3 m and from 11.7 m to 11.9 m	5.7	12.5
London Clay	Silty clay, stiff, moderately fissured and dark yellowish brown in colour	0.3+	12.8

GRADING

Mean for Deposit[15]				Depth below surface (m)[13]		Fines		Sand		Gravel	
	%	mm	%	From	To						
Gravel 50		+ 16	14	[16]**4.2	5.2	5	1	30	21	40	3
		- 16 + 4	36	**5.2	6.2	5	2	28	16	31	18
		- 4 + 1	17	**6.2	6.8	7	1	19	10	37	26
Sand 45		- 1 + $\frac{1}{4}$	27								
		- $\frac{1}{4}$ + 1/16	1								
Fines 5		- 1/16	5								

The numbered paragraphs below correspond with the annotations given on the specimen record above.

1. Borehole Registration Number
Each Industrial Minerals Assessment Unit (IMAU) borehole is identified by a Registration Number. This consists of two statements.
1) The number of the 1:25 000 sheet on which the borehole lies, for example TR 09.
2) The quarter of the 1:25 000 sheet on which the borehole lies and its number in a series for that quarter, for example NW 3.
Thus, the full Registration Number is TR 09 NW 3.

2. The National Grid Reference
All National Grid References in this publication lie within the 100-km squares TL, TM, TQ or TR unless otherwise stated. Grid references are given to eight figures, accurate to within 10 m for borehole locations. (In the text, six-figure grid references are used for more approximate locations, for example for farms.)

3. Location
The position of the borehole is generally referred to the nearest named locality on the 1:25 000 base map, and the resource block in which it lies is stated.

4. Surface level
The surface level at the borehole site is given in metres and feet above Ordnance Datum. All measurements were made in feet and subsequently converted to metres. An asterisk indicates that the surface level has been estimated.

5. Groundwater conditions
If groundwater was present the level at which it was encountered is normally given (in metres above Ordnance Datum).

6. Type of drill and date of drilling
Shell and auger rigs were used in this survey. The external diameter of the casing, and the month and year of completion of the borehole are stated.

7. Overburden, mineral, waste and bedrock
Mineral is sand and gravel, which, as part of a deposit, falls within the arbitrary definition of potentially workable material (see p. 1). Bedrock is the 'formation', 'country rock' or 'rock head' below which potentially workable sand and gravel will not be found.
  Waste is any material other than bedrock or mineral. Where waste occurs between the surface and mineral it is classified as overburden.

8. Thickness and depth
All measurements were made in metres. The thicknesses of beds and the depth from the surface of their bases have been recorded to the nearest 0.1 m.

9. The plus sign (+) indicates that the base of the deposit was not reached during drilling.

10. Geological classification
The geological classification (p. 3) is given whenever possible.

11. Lithological description
When sand and gravel is recorded a general description based on the mean grading characteristics (for details see Appendix C) is followed by more detailed particulars. The description of other rocks is based on visual examination in the field.

12. Rock colours were established in the field from the Rock Color Chart of the Geological Society of America.

13. Sampling
A continuous series of bulk samples is taken throughout the thickness of sand and gravel. A new sample is commenced whenever there is an appreciable lithological change within the sand and gravel or for every 1 m of depth.

14. Grading results
Exceptionally the results of the grading of a sample or horizon may not be available. No attempt has been made to estimate the probable grading of such samples.

15. Mean grading
The grading of the full thickness of the mineral horizon identified in the log is the mean of the individual sample gradings weighted by the thicknesses represented, if these vary. The classification used is shown in Table 4. Fully representative sampling of sand and gravel is difficult to achieve particularly where groundwater levels are high. Comparison between boreholes and adjacent exposures suggests that in borehole samples the proportion of sand may be higher and the proportion of fines and coarse gravel (+16 mm) may be lower.

16. A double asterisk indicates that the sample was bailed.

APPENDIX E: BOREHOLES USED IN THE ASSESSMENT OF RESOURCES

1. Industrial Minerals Assessment Unit and Engineering Geology Unit Boreholes

Borehole No. by sheet quadrant	Grid Reference	Page No.	Borehole No. by sheet quadrant	Grid Reference	Page No.
TL 80 NE			TM 00 NW		
49	8778 0505	28	17	0242 0537	50
TL 80 SE			TM 00 SW		
16	8914 0352	28	1	0006 0291	51
17	8921 0354	29	2	0020 0426	52
TL 90 NW			3	0182 0122	52
3	9424 0582	30	4	0241 0364	53
4	9077 0640	31	5	0131 0230	53
5	9203 0629	32	TQ 89 NE		
TL 90 NE			10	8847 9843	54
6	9971 0657	33	12	8645 9789	54
7	9872 0581	33	19*	8712 9659	55
8	9977 0844	34	23*	8782 9710	55
9	9954 0572	35	24*	8526 9656	56
TL 90 SW			TQ 99 NW		
1	9253 0031	36	25	9420 9653	57
TL 90 SE			26	9448 9946	58
1	9664 0035	37	27	9181 9891	59
2	9816 0194	38	29	9466 9787	60
3	9886 0183	39	31*	9447 9585	61
4	9892 0302	40	32*	9175 9721	61
5	9730 0269	41	40	9323 9901	62
6	9918 0443	42	TQ 99 NE		
7	9813 0408	43	15	9731 9632	63
8	9844 0030	43	16	9629 9617	64
9	9526 0185	44	17	9544 9832	65
10	9665 0412	45	18	9563 9921	66
TM 00 NW			19	9909 9771	66
9	0071 0854	46	20	9659 9815	67
10	0224 0799	46	21	9724 9822	68
11	0219 0626	47	22	9968 9666	69
12	0026 0762	47	23	9937 9932	70
13	0062 0547	48	24	9773 9986	70
14	0062 0566	48	25	9583 9692	71
15	0036 0673	49	26	9855 9668	72
16	0156 0745	49	27	9583 9952	73

*E.G.U. borehole

APPENDIX E: continued

Borehole No. by sheet quadrant	Grid Reference	Page No.
TR 09 NW		
2	0153 9762	74
3	0155 9974	75
5	0181 9589	76
6	0285 9650	76
8*	0105 9552	77
9*	0322 9775	78
10*	0307 9995	79

* E.G.U. Borehole

Detailed records may be consulted at the appropriate offices of the Institute, upon application to the Head, Industrial Minerals Assessment Unit.

2. Other boreholes

a. Dutch probes

Borehole No.	Grid Reference
TQ 99 NE	
28	9937 9558
29	9857 9808
30	9822 9574
TR 09 NW	
11	0051 9929
12	0232 9845
13	0203 9702
14	0075 9782
15	0086 9622

b. Hydrogeological Department records

Borehole No.	Grid Reference
242/3	0052 9933
242/23	0138 9905
258/9b	0472 9907
258/53	8922 9878
259/8b	9996 9646
259/11	9914 9831

c. Field Staff records

Borehole No.	Grid Reference
TL 90 SE	
11	971 013
12	992 046
TM 00 NW	
18	002 085
TM 00 SW	
6	023 045
7	029 002
8	017 002
TQ 99 NE	
31	969 993

d. Confidential records

7 boreholes

APPENDIX F: SUMMARY OF INFORMATION FROM INDUSTRIAL MINERALS ASSESSMENT UNIT AND ENGINEERING GEOLOGY UNIT BOREHOLES

Block A

Borehole No.	Thickness of mineral proved (m)	Thickness of drift (m)	Mean grading		
			Fines	Sand	Gravel
TL 90 NE					
6	0	5.8	-	-	-
7	3.1	12.1	6	55	39
8	0	1.6	-	-	-
9	5.8	5.8	4	35	61
TL 90 SE					
2	3.4	3.5	9	58	33
3	2.5	3.2	14	57	29
4	5.2	5.4	14	70	16
5	3.5	4.2	7	47	46
6	1.7	2.3	15	65	20
7	0	10.4	-	-	-
TM 00 NW					
9	0	3.8	-	-	-
10	0	1.9	-	-	-
12	3.2	3.5	9	44	47
14	0	0.9	-	-	-
15	0	2.4	-	-	-
16	1.2	2.1	7	87	6
TM 00 SW					
1	2.1	2.6	4	41	55
2	1.1	1.6	9	83	8
5	0	1.5	-	-	-

Block B

Borehole No.	Thickness of mineral proved (m)	Thickness of drift (m)	Mean grading		
			Fines	Sand	Gravel
TL 80 NE					
49	0	2.8	-	-	-
TL 80 SE					
16	0	2.0	-	-	-
17	0	1.8	-	-	-
TL 90 NW					
3	2.4	3.5	4	38	58
4	0	5.0	-	-	-
5	2.0	3.3	2	31	67
TL 90 SW					
1	0	13.0	-	-	-
TL 90 SE					
1	10.0	10.5	12	56	32
9	1.3	4.3	26	67	7
10	1.3	2.6	16	66	18
TQ 89 NE					
10	0	8.2	-	-	-
12	0	5.8	-	-	-
19	0	7.2	-	-	-
23	0	6.0	-	-	-
24	0	7.0	-	-	-
TQ 99 NW					
25	3.3	4.5	4	46	50
26	1.6	2.3	17	57	26
27	0	7.8	-	-	-
29	3.0	3.5	12	68	20
31	0	12.7	-	-	-
32	0	2.0	-	-	-
40	1.3	1.7	16	50	34
TQ 99 NE					
15	3.0	5.2	5	44	51
16	5.2	6.5	31	66	3
17	6.6	7.4	5	41	54
18	0	3.0	-	-	-
20	2.2	4.0+	13	30	57
24	0	7.9+	-	-	-
25	5.1	10.8	8	45	47
27	2.4	6.2	3	33	64

Block C

Borehole No.	Thickness of mineral proved (m)	Thickness of drift (m)	Mean grading		
			Fines	Sand	Gravel
TL 90 SE					
8	0	2.8	-	-	-
TM 00 NW					
11	0	4.3	-	-	-
13	0	5.9	-	-	-
17	0	12.0	-	-	-
TM 00 SW					
3	0	7.4	-	-	-
4	0	9.1	-	-	-
TQ 99 NE					
19	0	3.5	-	-	-
21	5.3	12.4	4	40	56
22	8.3	20.5	20	75	5
23	0	3.9	-	-	-
26	0	23.2	-	-	-
TR 09 NW					
2	0	16.2	-	-	-
3	2.6	12.5	5	45	50
5	0	21.3	-	-	-
6	0	15.0+	-	-	-
8	6.8	21.3	Grading results not available		
9	8.2	20.0	Grading results not available		
10	0	16.1	-	-	-

TL 80 NE 49	8778 0505	Mundon, Essex	Block B

Surface level(+1.9 m) +6.0 ft
Groundwater conditions not recorded
Shell and auger 203 mm diameter
March 1973

Waste 2.8 m
Bedrock 1.6 m+

LOG

Geological Classification	Lithology	Thickness m	Depth m
Soil and Marine or Estuarine Alluvium	Silty clay and clayey silt with carbonaceous material. Light bluish grey and light brown	2.8	2.8
London Clay	Silty clay. Light brown with patches of fine sand. Firm and highly fissured	1.6+	4.4

TL 80 SE 16	8914 0352	Mundon, Essex	Block B

Surface level(+4.2 m) +14.0 ft
Groundwater conditions not recorded
Shell and auger 203 mm diameter
February 1973

Waste 2.0 m
Bedrock 2.0 m+

LOG

Geological Classification	Lithology	Thickness m	Depth m
Soil and Marine or Estuarine Alluvium	Silty clay, stiff, with abundant plant remains and calcareous concretions near base. Mottled light brown and light bluish grey	2.0	2.0
London Clay	Silty clay with silty, fine sand laminae. Stiff, highly laminated and fissured. Light brown	2.0+	4.0

TL 80 SE 17   8921 0354   Mundon, Essex   Block B

Surface level(+3.4 m) +11.0 ft   Waste 1.8 m
Groundwater conditions not recorded   Bedrock 0.7 m+
Shell and auger 203 mm diameter
March 1973

LOG

Geological Classification	Lithology	Thickness m	Depth m
Soil and Marine or Estuarine Alluvium	Silty clay becoming clayey silt with depth. Soft with plant remains, mottled light brown and light bluish grey, grading into dusky blue and greenish grey	1.8	1.8
London Clay	Silty clay. Stiff and highly laminated. Mottled light brown and pale blue	0.7+	2.5

TL 90 NW 3   9424 0582   St Lawrence, Essex   Block B

Surface level(+3.8 m) +12.5 ft   Overburden 1.1 m
Water not struck   Mineral 2.4 m
Shell and auger 203 mm diameter   Bedrock 1.5 m+
March 1973

LOG

Geological Classification	Lithology	Thickness m	Depth m
Soil	Sandy silt with occasional gravel	0.3	0.3
River Loam (Brickearth)	Sandy silt, yellowish brown, laminated and friable, passing at 0.7 m into stiff, sandy, silty clay, light brown in colour and becoming sandier and gravelly with depth	0.8	1.1
River Terrace Sand and Gravel	Gravel Gravel: becoming coarser with depth with some cobbles near base of deposit. Mainly subangular and subrounded flints. Sand fine to coarse, but mainly fine to medium. Fairly high silt and clay content near top of mineral	2.4	3.5
London Clay	Silty clay, stiff and moderately fissured. Dark yellowish brown (weathered) to 4.5 m and dark grey (unweathered) below 4.5 m	1.5+	5.0

GRADING

Mean for Deposit				Depth below surface (m) From To		Bulk Samples Percentages Fines	Sand			Gravel	
	%	mm	%								
Gravel	58	+ 16	20	1.1	1.4	14	3	31	12	30	10
		- 16 + 4	38	** 1.4	1.7	6	2	35	12	34	11
				** 1.7	2.1	3	2	28	14	37	16
		- 4 + 1	14	** 2.1	2.7	1	1	17	16	41	24
Sand	38	- 1 + ¼	22	** 2.7	3.0	2	1	18	17	43	19
		- ¼ + 1/16	2	** 3.0	3.5	3	1	14	13	41	28
Fines	4	- 1/16	4								

Surface level(+6.9 m) +22.5 ft   Waste 5.0 m
Water not struck   Bedrock 1.2 m+
Shell and auger 203 mm diameter
April 1973

LOG

Geological Classification	Lithology	Thickness m	Depth m
Soil	Sandy silt	0.3	0.3
River Terrace Sand and Gravel	Sandy silt with rare gravel. Becoming more sandy and gravelly with depth. Roots fairly abundant	0.8	1.1

	'Clayey' sandy gravel
	   Gravel: fine to coarse, mainly sub-angular flints.
	   Sand: mainly fine to medium with some coarse grade material
		0.3	1.4

	Silty clay, becoming slightly sandy with depth. Race abundant near top. Stiff, light bluish grey, becoming greenish grey	1.8	3.2
	Very sandy silt with roots and abundant pebbles of vein-quartz. Sand mainly fine near top but becoming coarser and more abundant with depth	0.8	4.0

	Gravel
	   Gravel: coarse and fine with some cobbles predominantly angular to subangular flints.
	   Sand: coarse to fine mainly sub-angular
		1.0	5.0

London Clay	Silty clay with orange silt and fine sand along fissure planes near top of clay. Stiff, highly fissured and dark yellowish brown	1.2+	6.2

TL 90 NW 5   9203 0629   Osea Island, Essex   Block B

Surface level(+4.0 m) +13.0 ft   Overburden 1.3 m
Water not struck   Mineral 2.0 m
Shell and auger 203 mm diameter   Bedrock 0.6 m+
April 1973

LOG

Geological Classification	Lithology	Thickness m	Depth m
Made ground		1.1	1.1
River Terrace Sand and Gravel	Very sandy, slightly clayey silt, dark and moderate yellowish brown	0.2	1.3
	Gravel Gravel: coarse and fine, composed of angular to subangular flints and rounded reworked Tertiary pebbles of flint and quartzite. Sand coarse to fine, subangular to subrounded. Gravel becoming coarser and more abundant with depth	2.0	3.3
London Clay	Stiff, slightly silty highly fissured clay. Dark yellowish brown with blue reduced material along fissure planes	0.6+	3.9

GRADING

Mean for Deposit				Depth below surface (m)		Bulk Samples Percentages					
%	mm	%		From	To	Fines		Sand		Gravel	
Gravel 67	+ 16	31		**1.3	2.3	2	3	25	9	36	25
	- 16 + 4	36		**2.3	3.3	1	1	15	8	37	38
Sand 31	- 4 + 1	9									
	- 1 + ¼	20									
	- ¼ + 1/16	2									
Fines 2	- 1/16	2									

TL 90 NE 6   9971 0657   Bradwell, Essex   Block A

Surface level(+11.1 m) +36.5 ft   Waste 5.8 m
Water struck at +6.2 m   Bedrock 0.4 m+
Shell and auger 203 mm diameter
March 1973

LOG

Geological Classification	Lithology	Thickness m	Depth m
Soil	Sandy silt	0.3	0.3
Buried Channel Deposits	Intercalated bands of silt, clay and sand, generally 0.1 m to 0.2 m thick, but with thicker sandy silt bands from 0.3 m to 1.3 m and from 3.3 m to 5.5 m, and a sandy silty clay band from 1.6 m to 2.6 m. Carbonaceous throughout with race in places. Orange to yellowish brown	5.5	5.8
London Clay	Silty clay, stiff, very highly fissured and yellowish brown in colour	0.4+	6.2

TL 90 NE 7   9872 0581   Bradwell, Essex   Block A

Surface level(+15.5 m) +51.0 ft   Overburden 9.0 m
Water struck at +8.5 m   Mineral 3.1 m
Shell and auger 203 mm diameter   Bedrock 0.5 m+
March 1973

LOG

Geological Classification	Lithology	Thickness m	Depth m
Soil	Sandy silty clay	0.3	0.3
Head	Very sandy clay with occasional flint gravel. Yellowish brown and stiff	1.1	1.4
? River Terrace Sand and Gravel	Clayey sand with abundant fine to coarse gravel	0.3	1.7
Buried Channel Deposits	Sandy silty clay and clayey silt with sand lenses and bands. Occasional shell and reed bands. Mainly olive-grey in colour	7.3	9.0
	Sandy gravel Percentages of fine gravel and of fines increasing with depth, gravel consisting mainly of dark rounded flints. Sand coarse to fine but mainly of medium grade. Occasional fossil wood and shell fragments	3.1	12.1
London Clay	Silty clay. Firm to stiff, disturbed and dark yellowish brown	0.5+	12.6

33

TL 90 NE 8   9977 0844   Bradwell, Essex   Block A

Surface level(+2.7 m) +9.0 ft   Waste 1.6 m
Groundwater conditions not recorded   Bedrock 2.5 m+
Shell and auger 203 mm diameter
March 1973

LOG

Geological Classification	Lithology	Thickness m	Depth m
Soil and Marine or Estuarine Alluvium	Clayey, sandy, gravelly silt	0.6	0.6
	Sandy, clayey silt with rootlets. Soft to firm	1.0	1.6
London Clay	Silty clay, disturbed and weathered (yellowish brown) passing into undisturbed weathered material (dark yellowish brown) at 3.7 m depth. Firm, becoming stiff, with abundant 'race' near top	2.5+	4.1

TL 90 NE 9   9954 0572   Bradwell, Essex   Block A

Surface level (14.0 m) +46.0 ft   Mineral 5.8 m
Groundwater conditions not recorded   Bedrock 0.8 m+
Shell and auger 203 mm diameter
January 1974

LOG

Geological Classification	Lithology	Thickness m	Depth m
River Terrace Sand and Gravel	Gravel Gravel: coarse and fine, angular to subangular flints and rounded reworked Tertiary pebbles, rarely more than 30 mm median diameter. Sand mainly medium grade with some coarse grade	5.8	5.8
London Clay	Stiff, silty clay. Moderate to dark yellowish brown	0.8+	6.6

GRADING

Mean for Deposit				Depth below surface (m) From To		Fines	Bulk Samples Percentages Sand		Gravel	
	%	mm	%							
Gravel 61		+ 16	19	**0.1	0.9	5	2	42	17	27	7
		- 16 + 4	42	**0.9	1.7	4	3	21	19	43	10
				**1.7	2.3	2	3	19	8	38	30
		- 4 + 1	12	**2.3	3.2	3	3	16	8	45	25
Sand 35		- 1 + $\frac{1}{4}$	20	**3.2	4.5	4	2	9	9	49	27
		- $\frac{1}{4}$ + 1/16	3	**4.6	5.8	3	4	21	13	44	15
Fines 4		- 1/16	4								

Surface level (+43.5 m) +143.0 ft   Waste 13.0 m
Water not struck   Bedrock 2.1 m+
Shell and auger 203 mm diameter
March 1973

LOG

Geological Classification	Lithology	Thickness m	Depth m
Soil and River Terrace (? kame) Deposits	Sandy, silty clay, stiff with much organic material. Shells locally abundant. Yellowish brown becoming light brown and brownish grey with depth	1.1	1.1
Claygate Beds	Silty clay and sandy silty clay. Abundant 'race' near top. Organic debris common down to 5.0 m, and mica and selenite abundant throughout. Firm, becoming stiff, with colour varying from dark yellowish brown to dark yellowish orange	11.9	13.0
London Clay	Silty clay. Hard to stiff, finely laminated, greyish black	2.1+	15.1

TL 90 SE 1   9664 0035   Southminster, Essex   Block B

Surface level(+21.3 m) +70.0 ft   Overburden 0.5 m
Water struck at +15.3 m   Mineral 10.0 m
Shell and auger 203 mm diameter   Bedrock 2.6 m+
February 1973

LOG

Geological Classification	Lithology	Thickness m	Depth m
Soil	Sandy silt with rare gravel	0.5	0.5
River Terrace (and ? Glacial) Sand and Gravel	'Clayey' sandy gravel Becoming less silty but more gravelly with depth, being practically gravel-free between 1 m and 5 m depth. Elsewhere gravel coarse to fine angular to subangular flints and rounded reworked Tertiary pebbles. Sand coarse to fine but mainly medium grade, subangular to subrounded	10.0	10.5
London Clay	Stiff, silty, slightly sandy clay. Slightly carbonaceous near top. Dark yellowish brown (weathered) becoming dark olive-grey (unweathered) at 10.6 m depth	2.6+	13.1

GRADING

Mean for Deposit				Depth below surface (m)		Bulk Samples Percentages					
	%	mm	%	From	To	Fines	Sand			Gravel	
Gravel 32		+ 16	11	** 0.8	1.8	19	7	32	10	23	9
		- 16 + 4	21	** 1.8	2.9	24	16	58	1	1	0
				** 2.9	5.0	12	32	51	4	1	0
		- 4 + 1	6	** 5.0	6.0	Results not available					
Sand 56		- 1 + ¼	36	** 6.0	7.0	10	2	24	8	43	13
		- ¼ + 1/16	14	** 7.0	8.0	1	2	11	10	45	31
				** 8.0	9.0	8	5	20	4	37	26
Fines 12		- 1/16	12	** 9.0	10.5	Results not available					

TL 90 SE 2   9816 0194   Asheldham, Essex   Block A

Surface level(+15.8 m) +52.0 ft   Overburden 0.1 m
Water not struck   Mineral 3.4 m
Shell and auger 203 mm diameter   Bedrock 1.8 m+
March 1973

LOG

Geological Classification	Lithology	Thickness m	Depth m
Soil	Gravelly sand	0.2	0.2
River Terrace (and ? Glacial) Sand and Gravel	'Clayey' pebbly sand Gravel: mainly fine, subangular flints with some rounded pebbles. Sand mainly medium grade. Fines content low below 0.7 m depth	3.3	3.5
London Clay	Very silty clay, medium dark grey, stiff and moderately fissured	1.5+	5.3

GRADING

Mean for Deposit					Depth below surface (m)		Bulk Samples Percentages					
	%	mm	%		From	To	Fines	Sand			Gravel	
Gravel	33	+ 16	7		0.1	0.7	37	6	22	6	22	7
		- 16 + 4	26		0.7	1.8	5	5	42	11	33	4
					1.8	2.8	5	5	73	9	7	1
		- 4 + 1	8		2.8	3.5	5	6	83	4	2	0
Sand	58	- 1 + ¼	44									
		- ¼ + 1/16	6									
Fines	9	- 1/16	9									

Surface level (+20.5 m) +67.5 ft
Groundwater conditions not recorded
Shell and auger 203 mm diameter
March 1973

Overburden 0.7 m
Mineral 2.5 m
Bedrock 2.2 m+

LOG

Geological Classification	Lithology	Thickness m	Depth m
Soil	Silty sand with rare gravel	0.1	0.1
River Terrace Sand and Gravel	Very clayey silty fine sand	0.6	0.7
	'Clayey' sandy gravel Percentage of fines decreasing with depth. Coarse to fine flint gravel in coarse to fine, but predominantly medium-grade sand	2.5	3.2
London Clay	Silty clay, stiff, slightly fissured and dark grey	2.2+	5.4

GRADING

Mean for Deposit				Depth below surface (m)		Bulk Samples Percentages					
	%	mm	%	From	To	Fines		Sand		Gravel	
Gravel	29	+ 16	8	0.7	1.3	23	6	55	4	8	4
		- 16 + 4	21	**1.3	1.9	14	4	46	5	22	9
		- 4 + 1	10	**1.9	2.7	10	2	29	18	31	10
Sand	57	- 1 + $\frac{1}{4}$	43	**2.7	3.0	16	2	41	6	23	12
		- $\frac{1}{4}$ + 1/16	4	**3.0	3.2	7	5	72	11	5	0
Fines	14	- 1/16	14								

TL 90 SE 4   9892 0302   Tillingham, Essex   Block A

Surface level(+21.1 m) +69.0 ft   Overburden 0.2 m
Water struck at +18.4 m   Mineral 5.2 m
Shell and auger 203 mm diameter   Bedrock 0.6 m+
March 1973

LOG

Geological Classification	Lithology	Thickness m	Depth m
Soil		0.2	0.2
River Terrace (and ? Glacial) Sand and Gravel	'Clayey' pebbly sand, rather clayey near top of deposit and at 2.0 m depth Gravel: mainly rounded to sub-rounded flints, cobbles being common near base of deposit. Sand, coarse to fine but mainly of medium grade	5.2	5.4
London Clay	Silty clay, dark yellowish brown (weathered), firm and very disturbed, becoming unweathered at 5.6 m depth	0.6+	6.0

GRADING

Mean for Deposit

	%	mm	%
Gravel	16	+ 16	7
		- 16 + 4	9
Sand	70	- 4 + 1	5
		- 1 + ¼	57
		- ¼ + 1/16	8
Fines	14	- 1/16	14

Bulk Samples Percentages

Depth below surface (m) From	To	Fines		Sand		Gravel	
0.2	0.8	16	7	27	9	8	33
0.8	1.6	29	4	43	7	15	2
** 1.6	2.7	8	3	84	4	1	0
** 2.7	3.5	Results not available					
** 3.5	4.3	17	9	71	2	0	1
** 4.3	5.0	9	10	46	3	23	9
** 5.0	5.3	3	28	42	4	15	8

Surface level (+22.7 m) +74.5 ft
Water struck at +20.2 m
Shell and auger 203 mm diameter
March 1973

Overburden 0.7 m
Mineral 3.5 m
Bedrock 0.4 m+

LOG

Geological Classification	Lithology	Thickness m	Depth m
Soil	Gravelly, sandy, silty clay	0.2	0.2
Head	Sandy clay with occasional gravel. Stiff	0.5	0.7
River Terrace (and ? Glacial) Sand and Gravel	Sandy gravel becoming slightly less gravelly with depth, but with cobbles near base of deposit, otherwise fairly uniform composition throughout Gravel: rounded subangular and angular, coarse and fine	3.5	4.2
London Clay	Silty clay, dark yellowish brown	0.4+	4.6

GRADING

Mean for Deposit		
%	mm	%
Gravel 46	+ 16	14
	- 16 + 4	32
	- 4 + 1	11
Sand 47	- 1 + $\frac{1}{4}$	33
	- $\frac{1}{4}$ + 1/16	3
Fines 7	- 1/16	7

Depth below surface (m) From	To	Fines	Bulk Samples Percentages Sand			Gravel	
** 0.7	1.1	24	2	11	11	39	13
** 1.1	1.6	4	2	20	16	45	13
** 1.6	2.1	4	1	28	11	42	14
** 2.1	2.7	3	2	29	11	35	20
** 2.7	3.3	11	3	47	14	22	3
** 3.3	4.2	4	6	45	8	19	18

Surface level(+18.3 m) +60.0 ft   Overburden 0.6 m
Water not struck   Mineral 1.7 m
Shell and auger 203 mm diameter   Bedrock 1.2 m+
March 1973

LOG

Geological Classification	Lithology	Thickness m	Depth m
Soil and River Terrace Loam (Brickearth)	Clayey silt with sand	0.6	0.6
River Terrace (and ? Glacial) Sand and Gravel	'Clayey' pebbly sand Gravel content increasing and fines content decreasing with depth. Gravel: coarse and fine, dark rounded flints and light subangular flints. Sand coarse to fine but mainly of medium grade	1.7	2.3
London Clay	Silty clay, disturbed in the top 0.2 m, becoming undisturbed, firm to stiff and moderately fissured with depth. Moderate brown	1.2+	3.5

GRADING

Mean for Deposit				Depth below surface (m) From	To	Bulk Samples Percentages Fines		Sand		Gravel	
	%	mm	%								
Gravel 20		+ 16	10	0.7	1.6	18	7	62	8	5	0
		- 16 + 4	10	1.6	2.3	12	6	41	5	15	21
		- 4 + 1	7								
Sand	65	- 1 + $\frac{1}{4}$	52								
		- $\frac{1}{4}$ + 1/16	6								
Fines	15	- 1/16	15								

TL 90 SE 7   9813 0408   Tillingham, Essex   Block A

Surface level (+15.5 m) +51.0 ft   Waste 10.4 m
Water struck at +6.5 m   Bedrock 0.5 m+
Shell and auger 203 mm diameter
March 1973

LOG

Geological Classification	Lithology	Thickness m	Depth m
Soil	Gravelly, clayey, sandy silt	0.2	0.2
Head	Clayey, silty sand with a little gravel. Yellowish brown, becoming olive-grey and very calcareous at 1.0 m	2.2	2.4
Buried Channel Deposits	Clayey silt and silty clay with bands of more sandy aspect. Soft or occasionally firm, with abundant organic material and some shell bands	6.6	9.0
	Sandy gravel Gravel: mainly fine near top of deposit, but becoming coarser and more abundant with depth. Sand mainly of medium grade with a little coarse material	1.4	10.4
London Clay	Silty clay, dark yellowish brown	0.5+	10.9

TL 90 SE 8   9844 0030   Southminster, Essex   Block C

Surface level (+1.5 m) +5.0 ft   Waste 2.8 m
Water not struck   Bedrock 1.1 m+
Shell and auger 203 mm diameter
March 1973

LOG

Geological Classification	Lithology	Thickness m	Depth m
Soil	Clayey sandy silt	0.2	0.2
Marine or Estuarine Alluvium	Clayey silt and silty clay, organic in places with shell bands, and with rare gravel and 'race'	2.6	2.8
London Clay	Silty clay, firm with rootlets and 'race' nodules. Yellowish brown	1.1+	3.9

Surface level(+32.9 m) +108.0 ft	Overburden 1.4 m
Water not struck	Mineral 1.3 m
Shell and auger 203 mm diameter	Waste 1.6
February 1974	Bedrock 0.3 m+

LOG

Geological Classification	Lithology	Thickness m	Depth m
Soil	Clayey silt	0.4	0.4
River Terrace (? kame) Deposits	Sandy, silty clay with common carbonaceous material. Stiff, yellowish brown	1.0	1.4
	Very 'clayey' pebbly sand. Sand mainly fine, but becoming slightly coarser with depth	1.3	2.7
	Sandy, clayey silt	0.2	2.9
London Clay Head	Silty, sandy clay. Stiff, with abundant carbonaceous material and common 'race'	1.4	4.3
London Clay	Silty clay. Stiff, with rare 'race'	0.3+	4.6

GRADING

Mean for Deposit				Depth below surface (m)		Bulk Samples Percentages					
	%	mm	%	From	To	Fines	Sand			Gravel	
Gravel	7	+ 16	4	1.4	2.3	24	42	23	4	2	5
		- 16 + 4	3	2.3	2.7	28	27	32	5	5	3
		- 4 + 1	4								
Sand	67	- 1 + $\frac{1}{4}$	26								
		- $\frac{1}{4}$ + 1/16	37								
Fines	26	- 1/16	26								

TL 90 SE 10   9665 0412   St Lawrence, Essex   Block B

Surface level(+36.6 m) +120.0 ft   Overburden 1.2 m
Water not struck   Mineral 1.3 m
Shell and auger 203 mm diameter   Waste 0.1 m
February 1974   Bedrock 1.4m+

LOG

Geological Classification	Lithology	Thickness m	Depth m
Soil	Sandy, clayey silt	0.3	0.3
River Terrace (?kame) Deposits	Sandy, clayey silt and sandy, silty clay with occasional gravel becoming sandier with depth	1.1	1.4
	'Clayey' pebbly sand Gravel: coarse and fine, mainly angular flints with occasional rounded Tertiaries. Sand medium to fine	1.1	2.5
London Clay Head	Stiff, silty, sandy clay, bluish grey	0.1	2.6
London Clay	Stiff, silty, slightly sandy clay. Dark yellowish brown with blue streaks	1.4+	4.0

GRADING

Mean for Deposit				Depth below surface (m)		Bulk Samples Percentages					
	%	mm	%	From	To	Fines	Sand			Gravel	
Gravel	18	+ 16	5	**1.4	2.5	16	18	41	7	13	5
		- 16 + 4	13								
		- 4 + 1	7								
Sand	66	- 1 + ¼	41								
		- ¼ + 1/16	18								
Fines	16	- 1/16	16								

TM 00 NW 9   0071 0854   Bradwell, Essex   Block A

Surface level(+6.4 m) +21.0 ft   Waste 3.8 m
Water struck at +4.7 m   Bedrock 0.2 m+
Shell and auger 203 mm diameter
March 1973

LOG

Geological Classification	Lithology	Thickness m	Depth m
Soil	Slightly sandy silt	0.3	0.3
River Terrace Sand and Gravel	Sandy silt and clay, light yellowish brown becoming bluish grey and very gravelly with depth	2.0	2.3
London Clay Head	Gravelly silty clay, medium yellowish brown with abundant 'race', stiff	1.5	3.8
London Clay	Silty clay. Stiff, medium yellowish brown with high fissuring	0.2+	4.0

TM 00 NW 10   0224 0799   Bradwell, Essex   Block A

Surface level(+8.7 m) +28.5 ft   Waste 1.9 m
Water not struck   Bedrock 1.1 m+
Shell and auger 203 mm diameter
March 1973

LOG

Geological Classification	Lithology	Thickness m	Depth m
Soil	Slightly gravelly, sandy silt	0.4	0.4
River Terrace (and ? Glacial) Sand and Gravel	Sandy silt, becoming very silty sand with depth. Sand, mainly fine	1.5	1.9
London Clay	Slightly silty clay. Stiff, very highly fissured and mottled in shades of yellowish brown, contains selenite crystals	1.1+	3.0

TM 00 NW 11   0219 0626   Bradwell, Essex   Block C

Surface level(+1.7 m) +5.5 ft   Waste 4.3 m
Groundwater conditions not recorded   Bedrock 3.2 m+
Shell and auger 203 mm diameter
March 1973

LOG

Geological Classification	Lithology	Thickness m	Depth m
Soil	Slightly clayey silt	0.2	0.2
Marine or Estuarine Alluvium	Silty clay and clayey silt with peat band at 3.0 m. Soft beneath stiff top crust. Olive-grey in colour. Becoming sandy and gravelly near base of deposit	4.1	4.3
London Clay	Silty clay with abundant 'race' and selenite. Dark yellowish brown, stiff and very highly fissured	3.2+	7.5

TM 00 NW 12   0026 0762   Bradwell, Essex   Block A

Surface level(+9.5 m) +31.0 ft   Overburden 0.3 m
Water struck at 6.7 m   Mineral 3.2 m
Shell and auger 203 mm diameter   Bedrock 2.0 m+
March 1973

LOG

Geological Classification	Lithology	Thickness m	Depth m
Soil	Gravelly sandy clay	0.3	0.3
? River Terrace Sand and Gravel	Gravel Mainly rounded to subangular flints with a few cobbles, becoming sandier with less gravel and fines towards base. Sand of mainly medium grade throughout	3.2	3.5
London Clay	Silty clay, firm and dark yellowish brown (weathered) to 4.2 m, and firm, becoming stiff and slightly fissured (unweathered) below this depth	2.0+	5.5

GRADING

Mean for Deposit			Depth below surface (m) From	To	Bulk Samples Percentages Fines	Sand			Gravel	
%	mm	%								
Gravel 47	+ 16	16	** 0.3	1.5	17	2	13	15	42	11
	- 16 + 4	31	** 1.5	2.7	4	3	30	5	31	27
			** 2.7	3.5	3	2	63	8	15	9
	- 4 + 1	10								
Sand 44	- 1 + $\frac{1}{4}$	32								
	- $\frac{1}{4}$ + 1/16	2								
Fines 9	- 1/16	9								

TM 00 NW 13   0062 0547   Bradwell, Essex   Block C

Surface level(+2.8 m) +9.0 ft   Waste 5.9 m
Water not struck   Bedrock 0.4 m+
Shell and auger 203 mm diameter
March 1973

LOG

Geological Classification	Lithology	Thickness m	Depth m
Soil		0.2	0.2
Marine or Estuarine Alluvium	Clayey silt and silty clay. Mainly soft with firm to stiff layers. Reed remains common between 2.0 m and 4.3 m. Flint gravel in clayey silt from 4.3 m to 4.4 m	4.2	4.4
London Clay Head	Silty clay, stiff and mottled, with 'race' nodules and evidence of bioturbation	1.5	5.9
London Clay	Silty clay, stiff, highly fissured and of a uniform yellowish brown colour	0.4+	6.3

TM 00 NW 14   0062 0566   Bradwell, Essex   Block A

Surface level (+4.7 m) +15.5 ft   Waste 0.9 m
Water not struck   Bedrock 1.3 m+
Shell and auger 203 mm diameter
March 1973

LOG

Soil	Sandy, silty clay with gravel	0.4	0.4
London Clay Head	Silty clay with rare gravel. Firm, mottled and showing evidence of reworking and solifluction	0.5	0.9
London Clay	Silty clay with 'race' and pyrite. Stiff, highly fissured and cryoturbated	1.3+	2.2

TM 00 NW 15   0036 0673   Bradwell, Essex   Block A

Surface level(+11.5 m) +37.5 ft   Waste 2.4 m
Groundwater conditions not recorded   Bedrock 1.6 m+
Shell and auger 203 mm diameter
March 1973

LOG

Geological Classification	Lithology	Thickness m	Depth m
Made ground		1.8	1.8
?Buried Channel Deposits	Clayey sandy silt and sandy silty clay. Brown with green patches becoming greenish and bluish grey	0.6	2.4
London Clay	Silty clay, bioturbated and cryoturbated near top. Fissured, dark yellowish brown in colour	1.6+	4.0

TM 00 NW 16   0156 0745   Bradwell, Essex   Block A

Surface level(+9.9 m) +32.5 ft   Overburden 0.9 m
Water not struck   Mineral 1.2 m
Shell and auger 203 mm diameter   Bedrock 0.4 m+
March 1973

LOG

Geological Classification	Lithology	Thickness m	Depth m
Made ground		0.9	0.9
River Terrace (and ?Glacial) Sand and Gravel	Pebbly sand Gravel: mainly fine grade flint in medium grade sand	1.2	2.1
London Clay	Silty clay, yellowish brown and stiff	0.4+	2.5

GRADING

Mean for Deposit				Depth below surface (m)		Bulk Samples Percentages			
%	mm	%		From	To	Fines	Sand		Gravel
Gravel 6	+ 16	1		**1.0	2.1	7	4 77 6	5	1
	- 16 + 4	5							
	- 4 + 1	6							
Sand 87	- 1 + ¼	77							
	- ¼ + 1/16	4							
Fines 7	- 1/16	7							

49

Surface level(+2.4 m) +8.0 ft   Waste 12.0 m
Water struck at -6.9 m   Bedrock 0.9 m+
Shell and auger 203 mm diameter
February 1973

LOG

Geological Classification	Lithology	Thickness m	Depth m
Soil and Marine or Estuarine Alluvium	Clayey silt and silty clay with reed fragments and shell bands. Generally soft and orange-grey in colour	9.4	9.4
Beach Deposits	Very 'clayey' gravel Gravel: mainly rounded reworked Tertiary pebbles in fine to coarse sand	0.8	10.2
Marine or Estuarine Alluvium	Silty and sandy clay with rare gravel	1.8	12.0
London Clay	Silty clay, firm to stiff and yellowish brown	0.9+	12.9

Surface level(+18.6 m) +61.0 ft   Overburden 0.5 m
Groundwater conditions not recorded   Mineral 2.1 m
Shell and auger 203 mm diameter   Bedrock 0.4 m+
March 1973

LOG

Geological Classification	Lithology	Thickness m	Depth m
Soil	Sandy silt with a little gravel	0.3	0.3
River Terrace Sand and Gravel	Sandy gravelly silt, pale yellowish brown and dark yellowish orange	0.2	0.5
	Gravel Coarse gravel, mainly angular to subangular flints, with some rounded Tertiaries, fine gravel made up of roughly equal proportions of both types. Sand predominantly medium grade, subangular to subrounded	2.1	2.6
London Clay	Silty clay, stiff and highly fissured	0.4+	3.0

GRADING

Mean for Deposit			Depth below surface (m)		Bulk Samples Percentages					
%	mm	%	From	To	Fines	Sand		Gravel		
Gravel 55	+ 16	21	** 0.8	2.0	4	3	22	9	33	29
	- 16 + 4	34	** 2.0	2.6	4	3	33	18	36	6
Sand 41	- 4 + 1	12								
	- 1 + $\frac{1}{4}$	26								
	- $\frac{1}{4}$ + 1/16	3								
Fines 4	- 1/16	4								

TM 00 SW 2   0020 0426   Tillingham, Essex   Block A

Surface level (+10.3 m) +34.0 ft   Overburden 0.5 m
Water not struck   Mineral 1.1 m
Shell and auger 203 mm diameter   Bedrock 1.2 m+
March 1973

LOG

Geological Classification	Lithology	Thickness m	Depth m
Soil	Clayey sand with a little gravel	0.5	0.5
River Terrace Sand and Gravel	Pebbly sand Gravel: mainly fine and concentrated near base of deposit with occasional cobbles. Sand grade increasing with depth	1.1	1.6
London Clay	Silty clay with 'race' nodules. Dark yellowish brown, mottled green/grey	1.2+	2.8

GRADING

Mean for Deposit				Depth below surface (m) From	To	Bulk Samples Percentages Fines		Sand			Gravel
	%	mm	%								
Gravel	8	+ 16	2	** 0.5	1.0	15	21	60	3	1	0
		- 16 + 4	6	** 1.0	1.6	3	5	72	6	10	4
		- 4 + 1	5								
Sand	83	- 1 + ¼	66								
		- ¼ + 1/16	12								
Fines	9	- 1/16	9								

TM 00 SW 3   0182 0122   Dengie, Essex   Block C

Surface level (+2.0 m) +6.5 ft   Waste 7.4 m
Water not struck   Bedrock 1.0 m+
Shell and auger 203 mm diameter
March 1973

LOG

Geological Classification	Lithology	Thickness m	Depth m
Made ground		0.1	0.1
Marine or Estuarine Alluvium	Sand, silty and clay, generally soft, olive and light brownish grey with occasional shell and reed fragments	7.3	7.4
London Clay	Silty clay with 'race', weathered and disturbed	1.0+	8.2

TM 00 SW 4   0241 0364   Tillingham, Essex   Block C

Surface level(+2.4 m) +8.0 ft   Overburden 9.1 m
Water struck at 0.0 m   Bedrock 4.0 m+
Shell and auger 203 mm diameter
April 1973

LOG

Geological Classification	Lithology	Thickness m	Depth m
Soil	Slightly sandy, slightly clayey silt	0.2	0.2
Marine or Estuarine Alluvium	Clayey silt and silty clay with peat and shell bands. Olive-grey, frequently with blue and green mottling	8.9	9.1
London Clay	Slightly sandy, silty clay, sand content decreasing with depth, stiff and very highly fissured	4.0+	13.1

TM 00 SW 5   0131 0230   Tillingham, Essex   Block A

Surface level(+7.3 m) +24.0 ft   Waste 1.5 m
Water not struck   Bedrock 1.5 m+
Shell and auger 203 mm diameter
February 1974

LOG

Geological Classification	Lithology	Thickness m	Depth m
Made ground		0.5	0.5
River Terrace Loam (Brickearth)	Firm, clayey silt with occasional fine and coarse subangular flint gravel and medium and fine sand	0.8	1.3
River Terrace Sand and Gravel	Firm, very clayey silt with a little sand and abundant carbonaceous seeds	0.2	1.5
London Clay	Firm, very silty clay, cryoturbated with common carbonaceous remains and 'race' in top 1.1 m. Light to moderate brown with blue veining along root traces	1.5+	3.0

TQ 89 NE 10   8847 9843   Latchingdon, Essex   Block B

Surface level(+48.0 m) +157.5 ft   Waste 8.2 m
Groundwater conditions not recorded   Bedrock 1.9 m+
Shell and auger 203 mm diameter
March 1973

LOG

Geological Classification	Lithology	Thickness m	Depth m
Soil	Sandy clay with flints	0.4	0.4
River Terrace (? kame) Deposits	Very clayey sand with occasional gravel and organic debris	0.7	1.1
Head	Sandy silty clay with flint pebbles	1.1	2.2
Claygate Beds	Silty clay with sandy clayey silt band from 3.0 m to 3.5 m. Firm becoming stiff, and light brown becoming darker with depth. Unweathered below 6.2 m, where it is dark olive-grey	6.0	8.2
London Clay	Sandy silty clay. Stiff, moderately fissured and dark olive-grey	1.9+	10.1

TQ 89 NE 12   8645 9789   North Fambridge, Essex   Block B

Surface level(+1.8 m*) +6.0 ft*   Waste 5.8 m
Water struck at -2.6 m*   Bedrock 0.7 m+
Shell and auger 203 mm diameter
March 1973

LOG

Geological Classification	Lithology	Thickness m	Depth m
Soil and Marine or Estuarine Alluvium	Silty clay and clayey silt, peaty in places. Stiff near top, but becoming soft with depth	5.0	5.0
Buried Channel Deposits	Clayey silt with some coarse sand and flints and calcareous concretions	0.8	5.8
London Clay	Silty clay. Stiff, highly fissured and light brown	0.7+	6.5

TQ 89 NE 19   8712 9659   North Fambridge, Essex   Block B

Surface level (+1.6 m) +5.0 ft   Waste 7.2 m
Water struck at -5.4 m   Bedrock 1.0 m+
Shell and auger 203 mm diameter
April 1974

LOG

Geological Classification	Lithology	Thickness m	Depth m
Made ground		0.5	0.5
Marine or Estuarine Alluvium	Soft to stiff silty clay and clayey silt. Olive to bluish grey with peat band from 3.0 m to 3.5 m	6.5	7.0
Buried Channel Deposits	Clayey coarse sand	0.2	7.2
London Clay	Stiff, very silty clay with 'race' nodules and occasional sandy streaks. Dark yellowish brown with blue veining	1.0+	8.2

TQ 89 NE 23   8782 9710   North Fambridge, Essex   Block B

Surface level (+0.9 m) +3.0 ft   Waste 6.0 m
Groundwater conditions not recorded   Bedrock 1.0 m+
Shell and auger 203 mm diameter
March 1974

LOG

Geological Classification	Lithology	Thickness m	Depth m
Made ground		0.5	0.5
Marine or Estuarine Alluvium	Soft to very soft silty clay and silt with a little peat at 4.5 m. Olive-grey	4.6	5.1
London Clay Head	Soft to firm, very clayey silt with reed roots. Dark yellowish brown with blue mottling	0.9	6.0
London Clay	Fissured silty clay with rootlets. Dark yellowish brown with blue mottling	1.0+	7.0

Surface level (+1.2 m) +4.0 ft   Waste 7.0 m
Groundwater conditions not recorded   Bedrock 18.2 m+
Shell and auger 203 mm and 152 mm diameter
January 1974

LOG

Geological Classification	Lithology	Thickness m	Depth m
Soil		0.2	0.2
Marine or Estuarine Alluvium	Soft to firm sandy clayey silt and silty clay. Yellowish brown becoming dark olive-grey with occasional 'race'	4.9	5.1
London Clay Head	Stiff silty clay with common 'race' near top, and rare roots. Dark yellowish brown with blue veining	1.9	7.0
London Clay	Stiff silty clay with selenite. Dark yellowish brown becoming dark olive-grey at about 13.0 m. Thin cement-stone band at 8.5 m	18.2+	25.2

TQ 99 NW 25   9420 9653   Burnham, Essex   Block B

Surface level(+16.2 m) +53.0 ft   Overburden 1.2 m
Groundwater conditions not recorded   Mineral 3.3 m
Shell and auger 203 mm diameter   Bedrock 0.7 m+
March 1973

LOG

Geological Classification	Lithology	Thickness m	Depth m
Made ground		0.6	0.6
River Terrace Loam (Brickearth)	Clay with large angular flints. Stiff, moderately fissured and yellowish brown	0.6	1.2
River Terrace Sand and Gravel	Gravel	3.3	4.5
	Gravel: mainly fine, angular flint, with lesser amounts of coarse rounded Tertiary pebbles. Sand mainly fine at top of sample, but becoming coarser with depth. Very silty, clayey gravel from 3.9 m to 4.2 m		
London Clay	Silty clay, firm to stiff, orange-brown with rootlets	0.7+	5.2

GRADING

Mean for Deposit

	%	mm	%
Gravel	50	+ 16	13
		– 16 + 4	37
Sand	46	– 4 + 1	10
		– 1 + ¼	25
		– ¼ + 1/16	11
Fines	4	– 1/16	4

Depth below surface (m) From	To	Fines	Sand			Gravel	
1.4	2.0	16	42	25	3	9	5
** 2.0	2.8	0	3	24	8	41	24
** 2.8	3.3	0	4	30	13	40	13
** 3.3	3.9	3	2	16	13	54	12
** 4.2	4.5	1	2	37	16	39	5

Bulk Samples Percentages

TQ 99 NW 26   9448 9946   Southminster, Essex   Block B

Surface level (+29.2 m) +96.0 ft   Overburden 0.5 m
Groundwater conditions not recorded   Mineral 1.6 m
Shell and auger 203 mm diameter   Waste 0.2 m
March 1973   Bedrock 2.0 m+

LOG

Geological Classification	Lithology	Thickness m	Depth m
Made ground		0.3	0.3
River Terrace (? kame) Deposits	Gravelly silty sandy clay mottled orange and dark yellowish orange. Stiff	0.2	0.2
	'Clayey' sandy gravel Fines concentrated in lower part of deposit in clay-rich laminae. Gravel mainly fine angular flints concentrated in top of deposit	1.6	2.1
	Very sandy silty clay. Soft, finely laminated and carbonaceous. Dark yellowish orange, mottled light bluish grey	0.2	2.3
London Clay	Silty clay, firm to stiff, pale to light brown, with carbonaceous material	2.0+	4.3

GRADING

Mean for Deposit				Depth below surface (m) From	To	Fines	Sand			Gravel	
%		mm	%								
Gravel 26		+ 16	4	0.7	1.3	2	6	20	18	47	7
		- 16 + 4	22	1.3	1.6	26	17	48	2	4	3
Sand 57		- 4 + 1	9	**1.6	2.1	30	22	43	2	3	0
		- 1 + $\frac{1}{4}$	34								
		- $\frac{1}{4}$ + 1/16	14								
Fines 17		- 1/16	17								

Bulk Samples Percentages

58

Surface level (+44.7 m) +146.5 ft   Waste 7.8 m
Groundwater conditions not recorded   Bedrock 10.3 m+
Shell and auger 203 mm diameter
March 1973

LOG

Geological Classification	Lithology	Thickness m	Depth m
Soil and River Terrace (? kame) Deposits	Sandy clay with scattered angular to rounded flint gravel, firm with roots, dusky yellowish brown	0.3	0.3
Claygate Beds	Silty clay with sand laminae, in shades of brown in top 7.5 m (weathered) changing to dark olive grey (unweathered) in the lower portion. Occasional shell debris and organic matter. Stiff	7.5	7.8
London Clay	Silty, slightly sandy clay, finely laminated with shell bands. Stiff, highly fissured and dark olive-grey	10.3+	18.1

TQ 99 NW 29   9466 9787   Burnham, Essex   Block B

Surface level(+22.6 m) +74.0 ft   Overburden 0.3 m
Water not struck   Mineral 3.0 m
Shell and auger 203 mm diameter   Waste 0.2 m
March 1973   Bedrock 1.8 m+

LOG

Geological Classification	Lithology	Thickness m	Depth m
Soil	Sandy silt with a little gravel	0.3	0.3
River Terrace (?Glacial) Sand and Gravel	'Clayey' pebbly sand Percentage of gravel decreases with depth. Mainly fine with some coarse grade, subangular flints and rounded reworked Tertiary pebbles, occasionally shattered. Sand predominantly medium grade, subrounded. Deposit contains a thin, orange, very sandy silty clay band at about 1.5 m depth	3.0	3.3
Head	Silty slightly sandy clay, very disturbed with abundant carbonaceous material	0.2	3.5
London Clay	Silty, slightly sandy clay, slightly carbonaceous. Stiff very highly fissured, dark yellowish brown	1.8+	5.3

GRADING

Mean for Deposit				Depth below surface (m) From To		Bulk Samples Percentages		
%	mm	%				Fines	Sand	Gravel
Gravel 20	+ 16	6		** 0.5	1.5	8	2　33　13	29　15
	- 16 + 4	14		** 1.5	3.3	15	3　68　8	5　1
Sand 68	- 4 + 1	10						
	- 1 + ¼	55						
	- ¼ + 1/16	3						
Fines 12	- 1/16	12						

60

TQ 99 NW 31   9447 9585   Burnham, Essex   Block B

Surface level(+1.8 m) +6.0 ft   Waste 12.7 m
Water struck at +0.8 m   Bedrock 0.8 m+
Shell and auger 203 mm diameter
February 1974

LOG

Geological Classification	Lithology	Thickness m	Depth m
Soil		0.4	0.4
Marine or Estuarine Alluvium	Firm, becoming very soft clay, silt and fine sand, with shell band at 6.2 m. Dark grey	9.1	9.5
Buried Channel Deposits	Gravelly clay passing down at 11.6 m into clayey gravel with some sand. Gravel subangular to subrounded flints and black rounded reworked Tertiary pebbles	3.2	12.7
London Clay	Brown to grey silty clay, finely laminated and soliflucted near top	0.8+	13.5

TQ 99 NW 32   9175 9721   Althorne, Essex   Block B

Surface level(+1.5 m) +5.0 ft   Waste 2.0 m
Groundwater conditions not recorded   Bedrock 1.2 m+
Shell and auger 203 mm diameter
February 1974

LOG

Geological Classification	Lithology	Thickness m	Depth m
Soil		0.3	0.3
Marine or Estuarine Alluvium	Soft silty clay and laminated silt. Blue-grey and olive-brown with carbonaceous root remains	0.9	1.2
London Clay Head	Firm, clayey silt with scattered sand grains and flint pebbles	0.8	2.0
London Clay	Very silty clay with common mica flakes. Dark brown with blue veins	1.2+	3.2

Surface level(+38.0 m) +124.5 ft   Overburden 0.4 m
Groundwater conditions not recorded   Mineral 1.3 m
Shell and auger 203 mm diameter   Bedrock 1.4 m+
February 1974

LOG

Geological Classification	Lithology	Thickness m	Depth m
Soil		0.4	0.4
River Terrace (? kame) Deposits	'Clayey' sandy gravel Gravel: fine with some coarse, subangular to subrounded with about 30% black subrounded Tertiary reworked flints. Gravel becoming finer and less abundant with depth. Sand becoming finer with depth	1.3	1.7
London Clay	Firm to stiff very silty laminated clay with carbonaceous remains on bedding surfaces. Yellowish brown with blue reduced material along fissure planes	1.4+	3.1

GRADING

Mean for Deposit				Depth below surface (m)		Bulk Samples Percentages					
	%	mm	%	From	To	Fines	Sand			Gravel	
Gravel 34		+ 16	9	** 0.4	1.5	2	1	11	18	40	28
		- 16 + 4	25	** 1.5	1.7	18	7	32	15	22	6
Sand 50		- 4 + 1	15								
		- 1 + ¼	29								
		- ¼ + 1/16	6								
Fines 16		- 1/16	16								

Surface level (+1.8 m) +6.0 ft
Water struck at -2.6 m
Shell and auger 203 mm diameter
February 1973

Overburden 2.2 m
Mineral 3.0 m
Bedrock 5.3 m+

LOG

Geological Classification	Lithology	Thickness m	Depth m
Soil	Slightly clayey silt with a little gravel	0.3	0.3
Marine or Estuarine Alluvium	Silt, firm to soft, becoming gravelly and sandy near base. Dark yellowish brown becoming dark bluish grey	1.9	2.2
River Terrace Sand and Gravel	Gravel Mainly angular to subrounded flints, predominantly fine grade. Sand coarse to fine, subangular to subrounded	3.0	5.2
London Clay	Silty clay, stiff, highly fissured and dark yellowish brown to a depth of 9.5 m where it becomes unweathered, dark olive-grey and moderately fissured	5.3+	10.5

GRADING

Mean for Deposit

	%	mm	%
Gravel 51		+ 16	10
		- 16 + 4	41
		- 4 + 1	16
Sand 44		- 1 + $\frac{1}{4}$	27
		- $\frac{1}{4}$ + 1/16	1
Fines 5		- 1/16	5

Depth below surface (m) From	To	Fines	Bulk Samples Percentages Sand		Gravel	
** 2.4	3.4	7	1	32	10	37 13
** 3.4	4.4	4	1	30	16	40 9
** 4.4	5.2	3	1	16	24	47 9

TQ 99 NE 16   9629 9617   Burnham, Essex   Block B

Surface level(+4.6 m) +15.0 ft   Overburden 1.3 m
Water struck at +0.2 m   Mineral 5.2 m
Shell and auger 203 mm diameter   Bedrock 0.5 m+
February 1973

LOG

Geological Classification	Lithology	Thickness m	Depth m
Soil	Slightly sandy silt	0.3	0.3
Head Brickearth	Silt with a little gravel and with 'race' in lowest 0.1 m. Yellowish brown	0.8	1.1
River Terrace Sand and Gravel	Gravelly sandy silt	0.2	1.3
	Very 'clayey' sand Gravel: rare except near top of deposit. Below about 1.5 m, mineral very homogeneous. A band of sandy clayey silt occurs from 2.3 m to 2.4 m. Gravel mainly fine grade subangular to subrounded	5.2	6.5
London Clay	Slightly silty clay, stiff, dark yellowish brown and highly fissured	0.5+	7.0

GRADING

Mean for Deposit				Depth below surface (m)		Bulk Samples Percentages					
	%	mm	%	From	To	Fines	Sand			Gravel	
Gravel	3	+ 16	0	** 1.3	2.3	17	6	55	12	9	1
		- 16 + 4	3	** 2.4	3.4	35	8	55	1	1	0
				** 3.4	4.4	33	10	52	2	3	0
		- 4 + 1	4	** 4.4	5.4	30	10	56	2	2	0
Sand	66	- 1 + $\frac{1}{4}$	55	** 5.4	6.5	38	2	54	4	2	0
		- $\frac{1}{4}$ + 1/16	7								
Fines	31	- 1/16	31								

TQ 99 NE 17   9544 9832   Southminster, Essex   Block B

Surface level (+17.8 m) +58.5 ft   Overburden 0.8 m
Water not struck   Mineral 6.6 m
Shell and auger 203 mm diameter   Bedrock 0.6 m+
February 1973

LOG

Geological Classification	Lithology	Thickness m	Depth m
Soil	Clayey sandy gravelly silt	0.6	0.6
River Terrace Sand and Gravel	Clayey sand with gravel, gravel mainly coarse rounded Tertiary flints	0.2	0.8
	Gravel Gravel: coarse and fine, but mainly fine near base of deposit. Coarse gravel rounded Tertiary pebbles and subangular to subrounded flints, but fine gravel mainly angular flint and flint patina. Sand mainly medium grade, except near base of deposit where coarse sand predominates	6.6	7.4
London Clay	Silty clay. Firm, highly fissured, light brown (weathered) becoming unweathered at a depth of about 7.8 m	0.6+	8.0

GRADING

Mean for Deposit				Depth below surface (m)		Bulk Samples Percentages					
	%	mm	%	From	To	Fines	Sand			Gravel	
Gravel	54	+ 16	19	0.8	1.4	1	1	13	9	44	32
		- 16 + 4	35	**1.4	2.2	3	3	26	8	32	28
				**2.2	2.9	3	4	30	6	37	20
		- 4 + 1	11	**2.9	4.0	17	8	32	5	22	16
Sand	41	- 1 + ¼	26	**4.0	5.3	5	3	34	6	30	22
		- ¼ + 1/16	4	**5.3	6.4	1	4	32	18	35	10
				**6.4	7.4	2	1	8	22	50	17
Fines	5	- 1/16	5								

TQ 99 NE 18   9563 9921   Southminster, Essex   Block B

Surface level(+19.3 m) +63.5 ft   Waste 3.0 m
Water not struck   Bedrock 3.8 m+
Shell and auger 203 mm diameter
February 1973

LOG

Geological Classification	Lithology	Thickness m	Depth m
Soil	Sandy silt	0.5	0.5
Buried Channel Deposits	Very silty sandy clay with rare gravel. Light brown firm, becoming stiff	0.5	1.0
	Very sandy clayey silt with rare gravel. Light brown with light bluish grey patches. Gradational contact with overlying clay	2.5	3.0
London Clay	Silty clay, micaceous and slightly carbonaceous with lenses of fine orange sand near top. Yellowish brown, stiff and highly fissured	3.8+	6.8

TQ 99 NE 19   9909 9771   Burnham, Essex   Block C

Surface level(+1.1 m) +3.5 ft   Waste 3.5 m
Groundwater conditions   Bedrock 2.8 m+
Shell and auger 203 mm diameter
March 1973

LOG

Geological Classification	Lithology	Thickness m	Depth m
Made ground		1.4	1.4
Marine or Estuarine Alluvium	Sandy silt grading into silty sand with depth, with rare angular gravel. Dark olive-grey	2.1	3.5
London Clay	Silty clay with abundant 'race' near top, stiff, very highly fissured. Becoming slightly sandy with depth. Moderate brown	2.8+	6.3

Surface level(+4.7 m) +15.5 ft
Water not struck
Shell and auger 203 mm diameter
March 1973

Overburden 0.5 m
Mineral 2.2 m
Waste 1.3 m+

LOG

Geological Classification	Lithology	Thickness m	Depth m
Made ground		0.3	0.3
Head	Gravelly sandy silt. Orange	0.2	0.5

	'Clayey' gravel
	    Gravel: mainly fine angular flints
	       with some reworked Tertiary
	       pebbles. Sand coarse to fine,
	       but mainly of medium grade.
	       Deposit becoming more silty
	       towards base
		2.2	2.7

Buried Channel Deposits	Very silty clay, carbonaceous in top 1 m, yellowish brown	1.3+	4.0

GRADING

Mean for Deposit				Depth below surface (m) From	To	Bulk Samples Percentages Fines		Sand		Gravel	
	%	mm	%								
Gravel 57		+ 16	11	** 0.5	1.8	3	1	21	7	49	19
		- 16 + 4	46	** 1.8	2.7	27	6	18	7	42	0
Sand	30	- 4 + 1	7								
		- 1 + $\frac{1}{4}$	20								
		- $\frac{1}{4}$ + 1/16	3								
Fines	13	- 1/16	13								

Surface level (+1.2 m) +4.0 ft   Overburden 7.1 m
Water struck at -0.80 m   Mineral 5.3 m
Shell and auger 203 mm diameter   Bedrock 0.6 m+
March 1973

LOG

Geological Classification	Lithology	Thickness m	Depth m
Soil		0.2	0.2
Marine or Estuarine Alluvium	Clayey and sandy silt, generally soft, but firm in places. Olive and greenish grey	4.4	4.6
Buried Channel Deposits	Sandy and clayey silt. Firm, laminated with abundant 'race' and organic remains	2.5	7.1
	Gravel Percentage of coarse and fine gravel fairly constant throughout but becoming finer with depth. Mainly angular to subangular flints with some reworked Tertiary pebbles. Sand medium to coarse, subangular to subrounded	5.3	12.4
London Clay	Silty clay with selenite. Stiff and dark yellowish brown	0.6+	13.0

GRADING

Mean for Deposit				Depth below surface (m) From To		Bulk Samples Percentages Fines		Sand		Gravel	
	%	mm	%								
Gravel	56	+ 16	23	** 7.1	8.1	8	4	31	6	22	29
		- 16 + 4	33	** 8.1	9.1	5	3	26	8	27	31
				** 9.1	10.1	2	2	19	13	39	25
		- 4 + 1	11	** 10.1	11.1	3	1	28	11	41	16
Sand	40	- 1 + $\frac{1}{4}$	27	** 11.1	12.4	3	1	29	15	35	17
		- $\frac{1}{4}$ + 1/16	2								
Fines	4	- 1/16	4								

TQ 99 NE 22   9968 9666   Burnham, Essex   Block C

Surface level(+1.5 m) +5.0 ft   Overburden 12.2 m
Water struck at -10.4 m   Mineral 8.3 m
Shell and auger 203 mm diameter   Bedrock 1.0 m+
February 1973

LOG

Geological Classification	Lithology	Thickness m	Depth m
Soil	Sandy silty clay	0.7	0.7
Marine or Estuarine Alluvium	Silty clay and clayey sand and silt with occasional shell fragments and abundant organic debris	11.5	12.2
Buried Channel Deposits	'Clayey' pebbly sand Percentage of gravel negligible above 17.8 m depth. Below this level gravel proportion increases with depth and is abundant with scattered flint cobbles near base of deposit, where it consists of flint with some rounded quartzite pebbles. Sand mainly medium and fine	8.3	20.5
London Clay	Silty clay, firm with no fissuring and olive-grey in colour	1.0+	21.5

GRADING

Mean for Deposit				Depth below surface (m)		Bulk Samples Percentages					
	%	mm	%	From	To	Fines	Sand			Gravel	
Gravel 5		+ 16	2	**12.2	13.5	21	18	59	1	1	0
		- 16 + 4	3	**13.5	16.4	31	16	52	1	0	0
				**16.4	17.8	7	23	67	2	1	0
		- 4 + 1	3	**17.8	19.1	19	26	40	5	9	1
Sand 75		- 1 + ¼	53	**19.1	19.8	9	16	60	11	3	1
		- ¼ + 1/16	19	**19.8	20.5	9	11	34	9	12	25
Fines 20		- 1/16	20								

TQ 99 NE 23　　　　　　9937 9932　　　　　　Southminster, Essex　　　　　　Block C

Surface level (+1.7 m) +5.5 ft　　　　　　　　　　　　　Waste 3.9 m
Water struck at +0.2 m　　　　　　　　　　　　　　　　Bedrock 2.1 m+
Shell and auger 203 mm diameter
February 1973

LOG

Geological Classification	Lithology	Thickness m	Depth m
Soil and Marine or Estuarine Alluvium	Thin bands of sand, silt and clay, pale grey and brown	1.2	1.2
Buried Channel Deposits	Clayey, sandy silt and sand with occasional gravel	1.0	2.2
	Silty clay, stiff with 'race' nodules and rootlets. Yellowish brown and greenish grey	1.7	3.9
London Clay	Silty clay, stiff, moderately fissured and dark yellowish brown	2.1+	6.0

TQ 99 NE 24　　　　　　9773 9986　　　　　　Southminster, Essex　　　　　　Block B

Surface level (4.4 m) +14.5 ft　　　　　　　　　　　　Waste 7.9 m+
Groundwater conditions not recorded
Shell and auger 203 mm diameter
February 1973

LOG

Geological Classification	Lithology	Thickness m	Depth m
Made ground		1.0	1.0
Buried Channel Deposits	Clayey silt, becoming silty clay at 1.2 m depth. Occasional quartz and flint pebbles and abundant rootlets. Shell band from 2.5 m to 3.2 m depth. Pale yellowish brown, grading into light olive-brown, through olive-grey to bluish grey	6.9+	7.9

TQ 99 NE 25   9583 9692   Burnham, Essex   Block B

Surface level(+6.1 m) +20.0 ft   Overburden 5.7 m
Water struck at +0.5 m   Mineral 5.1 m
Shell and auger 203 mm diameter   Bedrock 1.0 m+
February 1973

LOG

Geological Classification	Lithology	Thickness m	Depth m
Made ground		0.7	0.7
Head Brickearth	Clayey sandy gravelly silt with 'race'. Mottled light brown and greenish grey	1.3	2.0
Buried Channel Deposits	Silty clay with 'race' nodules. Light olive-grey. Firm becoming stiff with depth. Pale yellowish brown immediately above mineral	3.7	5.7
	Gravel	5.1	10.8
	Gravel: coarse to fine, rounded Tertiary pebbles and angular to subangular flints. Sand, coarse to fine but mainly medium grade, subangular to subrounded flint and clear quartz. Occasional silty clay lumps. Proportions of fine angular flint gravel and coarse sand higher than average near base of deposit		
London Clay	Silty, slightly sandy clay. Dark yellowish brown (weathered) in top 0.1 m changing to olive-grey in unweathered material	1.0+	11.8

GRADING

Mean for Deposit				Depth below surface (m) From To		Bulk Samples Percentages Fines		Sand		Gravel	
	%	mm	%	From	To						
Gravel 47		+ 16	18	** 5.7	6.7	Results not available					
		- 16 + 4	29	** 6.7	7.7	7	1	42	13	29	8
				** 7.7	8.7	4	2	30	11	25	28
Sand 45		- 4 + 1	11	** 8.7	9.7	16	2	31	5	25	21
		- 1 + ¼	32	** 9.7	10.8	6	2	27	13	37	15
		- ¼ + 1/16	2								
Fines 8		- 1/16	8								

TQ 99 NE 26   9855 9668   Burnham, Essex   Block C

Surface level(+2.0 m) +6.5 ft   Waste 23.2 m
Water struck at -5.3 m   Bedrock 2.0 m+
Shell and auger 203 mm and 152 mm diameter
February 1973

LOG

Geological Classification	Lithology	Thickness m	Depth m
Soil	Slightly clayey silt	0.4	0.4
Marine or Estuarine Alluvium	Sandy and clayey silt and clay with shell band from 20.5 m to 22.5 m	22.1	22.5
Beach Deposits	Pebbly sand with shells	0.7	23.2
London Clay	Silty clay with selenite. Stiff, highly fissured, very dark yellowish brown	2.0+	25.2

TQ 99 NE 27   9583 9952   Southminster, Essex   Block B

Surface level(+20.7 m) +68.0 ft   Overburden 3.8 m
Water struck at +16.9 m   Mineral 2.4 m
Shell and auger 203 mm diameter   Bedrock 1.2 m+
February 1974

LOG

Geological Classification	Lithology	Thickness m	Depth m
Soil		0.2	0.2
River Terrace Sand and Gravel	Sandy silty clay, becoming more sandy with depth. Orange to yellowish brown in colour with abundant carbonaceous root remains in top half	3.6	3.8
	Gravel	2.4	6.2
	Gravel: coarse and fine, becoming coarser with depth and consisting of roughly equal proportions of black rounded reworked Tertiary pebbles and subangular to subrounded flints. Sand mainly medium to fine grade		
London Clay	Stiff very silty clay with common mica. Brown becoming dark olive-grey at 6.5 m	1.2+	7.4

GRADING

Mean for Deposit				Depth below surface (m) From	To	Bulk Samples Percentages Fines	Sand			Gravel	
	%	mm	%								
Gravel	64	+ 16	33	** 4.0	5.0	5	10	18	6	34	27
		- 16 + 4	31	** 5.0	5.6	3	12	16	4	29	36
				** 5.6	6.2	1	7	22	4	27	39
		- 4 + 1	5								
Sand	33	- 1 + ¼	18								
		- ¼ + 1/16	10								
Fines	3	- 1/16	3								

Surface level(+2.0 m) +6.5 ft   Waste 16.2 m
Water struck at -7.9 m   Bedrock 0.8 m+
Shell and auger 203 mm diameter
February 1973

LOG

Geological Classification	Lithology	Thickness m	Depth m
Soil	Gravelly silty clay with shells	0.2	0.2
Marine or Estuarine Alluvium	Intercalated bands of silty fine sand, silt and clay, brown near the weathered surface grading into bluish grey and dark olive-grey; abundant shell debris	16.0	16.2
London Clay	Silty clay, disturbed and weathered. Greyish brown, firm at top becoming stiff with depth	0.8+	17.0

Surface level(+2.1 m) +7.0 ft   Overburden 4.2 m
Water struck at -2.1 m   Mineral 2.6 m
Shell and auger 203 mm diameter   Waste 5.7 m
February 1973   Bedrock 0.3 m+

LOG

Geological Classification	Lithology	Thickness m	Depth m
Soil	Clayey silt with broken shells	0.2	0.2
Marine or Estuarine Alluvium	Sandy clay and silt with occasional shells and organic debris. Mainly medium bluish grey	4.0	4.2
Buried Channel Deposits	Gravel Gravel: becoming coarser with depth, being mainly fine grade near top of deposit, but having almost equal proportions of fine and coarse gravel	2.6	6.8
	Silty clay and clayey silt with occasional reeds and shell fragments. Soft to stiff, dark bluish grey with some olive-grey and yellowish brown bands from 10.2 m to 11.3 m and from 11.7 m to 11.9 m	5.7	12.5
London Clay	Silty clay, stiff, moderately fissured and dark yellowish brown in colour	0.3+	12.8

GRADING

Mean for Deposit				Depth below surface (m) From To		Fines	Bulk Samples Percentages Sand		Gravel	
	%	mm	%							
Gravel	50	+ 16	14	** 4.2	5.2	5	1	30	21	40	3
		- 16 + 4	36	** 5.2	6.2	5	2	28	16	31	18
				** 6.2	6.8	7	1	19	10	37	26
		- 4 + 1	17								
Sand	45	- 1 + $\frac{1}{4}$	27								
		- $\frac{1}{4}$ + 1/16	1								
Fines	5	- 1/16	5								

TR 09 NW 5   0181 9589   Burnham, Essex   Block C

Surface level(+1.8 m) +6.0 ft   Waste 21.3 m
Water struck at -0.5 m   Bedrock 0.3 m+
Shell and auger 203 mm and 152 mm diameter
March 1973

LOG

Geological Classification	Lithology	Thickness m	Depth m
Soil	Slightly clayey silt	0.1	0.1
Marine or Estuarine Alluvium	Sandy and clayey silt and silty clay with shelly gravelly sand from 14.4 m to 15.4 m overlain by a 0.4 m thick shell band	20.3	20.4
Buried Channel Deposits	Gravel Mainly fine gravel in fine to coarse sand, coarse gravel being found only near base of deposit	0.9	21.3
London Clay	Silty clay with selenite, stiff dark yellowish brown in colour and highly fissured	0.3+	21.6

TR 09 NW 6   0285 9650   Burnham, Essex   Block C

Surface level(+1.3 m) +4.5 ft   Waste 15.0 m+
Water struck at -1.1 m
Shell and auger 203 mm diameter
March 1973

LOG

Geological Classification	Lithology	Thickness m	Depth m
Soil	Slightly clayey silt	0.1	0.1
Marine or Estuarine Alluvium	Sandy and clayey silt and silty clay, pale to dark olive-grey in colour	13.7	13.8
Buried Channel Deposits	Very silty sand with rare gravel	1.2+	15.0

TR 09 NW 8   0105 9552   Burnham, Essex   Block C

Surface level(+1.7 m) +5.5 ft   Overburden 14.5 m
Groundwater conditions not recorded   Mineral 6.8 m
Shell and auger 203 mm diameter   Bedrock 1.0 m+
March 1974

LOG

Geological Classification	Lithology	Thickness m	Depth m
Made ground		0.6	0.6
Marine or Estuarine Alluvium	Firm to very soft silty fine sand and clayey silt with occasional peat partings and reeds and scattered shell fragments. Bluish to olive-grey in colour	13.9	14.5
Buried Channel Deposits	Fine to coarse gravel in a fine sand matrix. Flints becoming finer and sand content increasing with depth. Brown becoming white or grey at about 18.7 m. Cobbles and common shell fragments near base of deposit	6.8	21.3
London Clay	Firm, brown weathered silty clay, rapidly becoming stiff, fissured unweathered silty clay	1.0+	22.3

Grading results not available

TR 09 NW 9 　　　　　0322 9775 　　　　　Southminster, Essex 　　　　　Block C

Surface level(-0.2 m) -0.5 ft 　　　　　　　　　　　　　Overburden 11.8 m
Water struck at -9.1 m 　　　　　　　　　　　　　　　　Mineral 8.2 m
Shell and auger 203 mm and 152 mm diameter 　　　　Bedrock 2.0 m+
March 1974

LOG

Geological Classification	Lithology	Thickness m	Depth m
Made ground		0.3	0.3
Marine or Estuarine Alluvium	Soft to very soft clayey sandy silt and silty fine sand, olive or dark grey with abundant shells between 4.8 m and 7.4 m	11.5	11.8
Buried Channel Deposits	Gravelly sand, becoming sandy gravel with depth, with abundant shell material. Gravel grade increasing with depth	3.8	15.6
	Gravelly sand, rapidly becoming sandy gravel, but with rare gravel between 18.0 m and 19.0 m. Gravel coarse to fine, mainly dark, rounded to subrounded Tertiary reworked flints with occasional subangular flints. Sand medium to fine grade	4.4	20.0
London Clay	Firm to stiff silty clay. Disturbed near top but rapidly showing signs of fissuring	2.0+	22.0

Grading results not available

Surface level(+0.7 m) +2.5 ft
Groundwater conditions not recorded
Shell and auger 203 mm diameter
February 1974

Waste 16.1 m
Bedrock 4.2 m+

LOG

Geological Classification	Lithology	Thickness m	Depth m
Marine or Estuarine Alluvium	Soft clayey sandy silt. Dark grey and bluish grey with occasional reed fragments and some shells near base of deposit	14.5	14.5
Buried Channel Deposits	Sand with rare fine gravel with some calcareous cement	1.2	15.7
London Clay Head	Reddish brown very clayey silt. Firm with some fine sand and mica and occasional root traces	0.4	16.1
London Clay	Stiff silty clay. Highly fissured and dark yellowish brown in colour. Becoming unweathered at 20 m	4.2+	20.3

Surface level(21.5 m*) 70.5 ft*   Mineral 3.6 m+
Water standing at 16.9 m*
Exposure
June 1973

LOG

Geological Classification	Lithology	Thickness m	Depth m
River Terrace (and ? Glacial) Sand and Gravel	'Clayey' pebbly sand Clay throughout top 1 m of deposit. Discreet silty clayey bands in lowest 1 m. Gravel coarse and fine, mainly angular and rounded flints with less than 5% vein-quartz and sedimentary rock fragments. Trough cross- bedded throughout	3.6+	3.6

Lateral variations

Exposure too limited to record lateral variation

GRADING

Mean for Deposit

	%	mm	%
Gravel	17	+ 16	6
		- 16 + 4	11
		- 4 + 1	10
Sand	69	- 1 + $\frac{1}{4}$	52
		- $\frac{1}{4}$ + 1/16	7
Fines	14	- 1/16	14

Surface level(16.8 m*) 55.0 ft*   Overburden 0.9 m
Dry pit   Mineral 4.4 m+
Exposure
June 1973

LOG

Geological Classification	Lithology	Thickness m	Depth m
Made ground		0.9	0.9
River Terrace (and ? Glacial) Sand and Gravel	Sandy gravel Gravel concentrated between about 2 m and 4 m depth, with pebbly sand above and below. More sandy horizons are trough cross-bedded and poorly cemented, cementation being fairly good in the gravels. Gravel coarse and fine angular and rounded flints with less than 5% quartz and sedimentary rock fragments	4.4+	5.3

Lateral variations

No significant lateral variation noted

GRADING

Mean for Deposit

	%	mm	%
Gravel	35	+ 16	15
		- 16 + 4	20
Sand	61	- 4 + 1	7
		- 1 + $\frac{1}{4}$	51
		- $\frac{1}{4}$ + 1/16	3
Fines	4	- 1/16	4

Surface level(20.0 m*) 65.5 ft*   Overburden 0.3 m
Water level not recorded   Mineral 4.7 m+
Exposure
June 1973

LOG

Geological Classification	Lithology	Thickness m	Depth m
Soil		0.3	0.3
River Terrace (and ? Glacial) Sand and Gravel	Pebbly sand Gravel concentrated in top 0.5 m. Elsewhere, mainly fine gravel limited to base of trough cross-bedded sets and lying along planes of foresets	4.7+	5.0

Lateral variations

Deposit becomes more gravelly to north

GRADING

Mean for Deposit

	%	mm	%
Gravel	5	+ 16	1
		− 16 + 4	4
Sand	93	− 4 + 1	2
		− 1 + $\frac{1}{4}$	66
		− $\frac{1}{4}$ + 1/16	25
Fines	2	− 1/16	2

Surface level(21.5 m*) 70.5 ft*   Overburden 0.9 m
Water standing at 14.8 m*   Mineral 5.8 m+
Exposure
June 1973

LOG

Geological Classification	Lithology	Thickness m	Depth m
Soil	Sandy silt	0.2	0.2
River Terrace Loam (Brick-earth)	Sandy silt with roots	0.7	0.9
River Terrace (and ? Glacial) Sand and Gravel	'Clayey' sandy gravel Trough cross-bedded with small gravel-filled channel at about 2 m depth. Deposit fining upwards above erosional base of channel and from lowest exposed horizon, becoming slightly gravelly sand below the channel. Gravel mainly coarse angular to subangular and rounded flints, with less than 5% vein-quartz and sedimentary rock fragments	5.8+	6.7

Lateral variations

Deposit becomes progressively more
    silty to the west, having a fines
    content of about 30%, 70 m to the
    west, where the sand is micro-
    trough cross-laminated. There is
    little apparent lateral variation in
    composition in a north-south
    direction. Local palaeocurrent
    directions are approximately towards
    the north-north-east

GRADING

Mean for Deposit

	%	mm	%
Gravel	33	+ 16	28
		- 16 + 4	5
Sand	52	- 4 + 1	2
		1 + ¼	37
		- ¼ + 1/16	13
Fines	15	- 1/16	15

Surface level (18.3 m*) 60.0 ft*   Overburden 2.9 m
Water standing at 12.0 m*   Mineral 2.7 m+
Exposure
June 1973

LOG

Geological Classification	Lithology	Thickness m	Depth m
Soil		0.2	0.2
River Terrace Loam (Brick-earth)	Micaceous silt with rare gravel and abundant roots	2.7	2.9
River Terrace (and ? Glacial) Sand and Gravel	Sand Gravel: mainly of fine grade, concentrated in top 0.5 m. Trough cross-bedded with local palaeo-current direction towards the north-east	2.7+	5.6

Lateral variations

Deposit becomes more gravelly towards the east

GRADING

Mean for Deposit

	%	mm	%
Gravel	4	+ 16	1
		- 16 + 4	3
Sand	94	- 4 + 1	2
		- 1 + $\frac{1}{4}$	84
		- $\frac{1}{4}$ + 1/16	8
Fines	2	- 1/16	2

APPENDIX H: RESISTIVITY SURVEY RESULTS

The author acknowledges the work of
Mr M. Sarginson who supplied information upon
which this appendix is based.

In order to determine more accurately the
shape, location and trend of the Burnham Channel,
whose presence had been indicated by silts and
clays encountered in boreholes TQ 99 NE 20,
NE 24 and NE 25 and by several auger holes in
the area to the east and south-east of Southminster,
a resistivity survey was carried out in the vicinity
of Dammer Wick Farm [9626 9692].

An A.B.E.M. Terrameter in its A.C. version
was used in this survey, which included one
horizontal traverse between a point to the west of
Dammer Wick Farm [9615 9692] and the railway

bridge [9540 9689], the line of traverse passing
close to borehole TQ 99 NE 25. The borehole was
used to confirm the nature of the lithologies
producing lateral variations deduced from the
geophysics. Using the table of apparent resistivity ranges for lithologies in the area (Fig. 11),
the near-surface changes in lithology can be
accurately plotted across the line of traverse.

The apparent resistivity values Qa measured
in ohm metres, obtained at 5-m intervals, are
plotted against the position of the centre of the
electrode configuration in Figure 9, while
Figure 10 shows the estimated near-surface
geology. Once the lithology is known, then an
approximate thickness of the lithological units
can be derived from the position of the measured
apparent resistivity value in the known range.

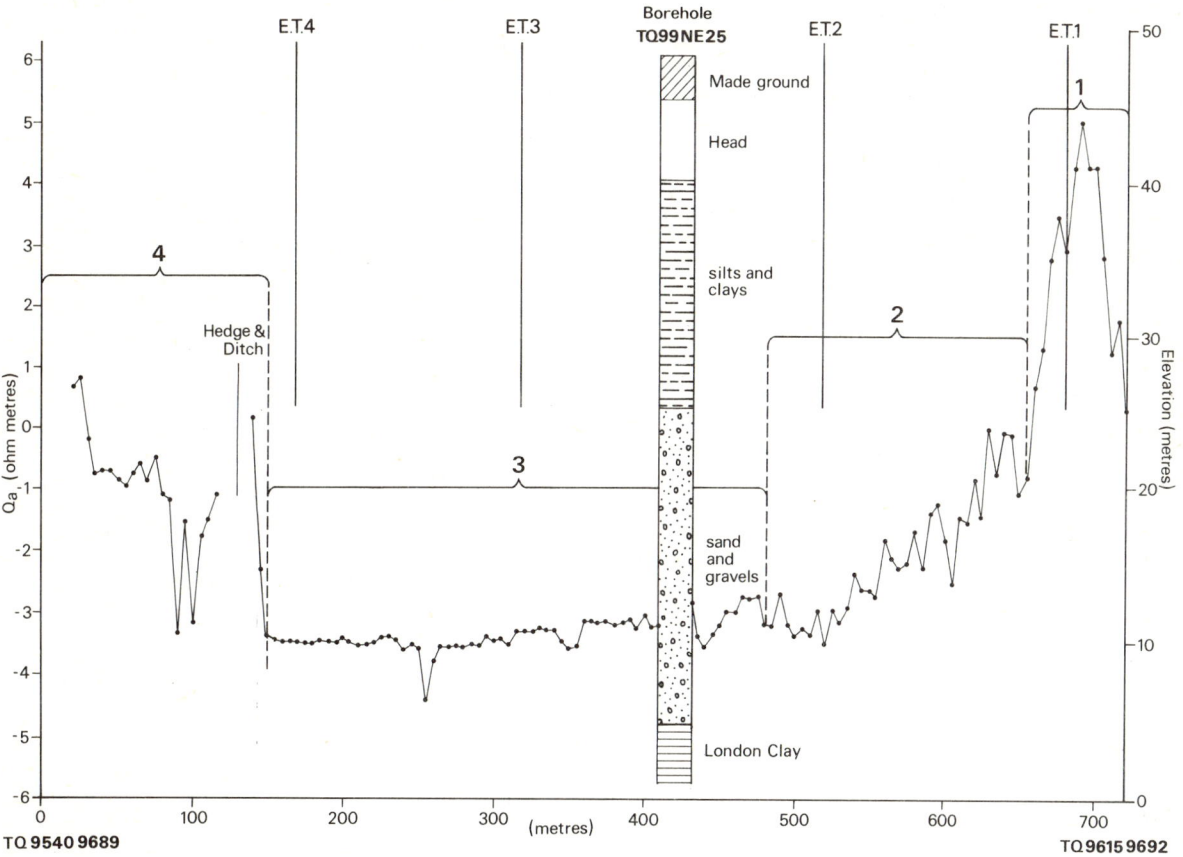

Fig. 9   Terrameter traverse in the region of the Burnham Buried Channel

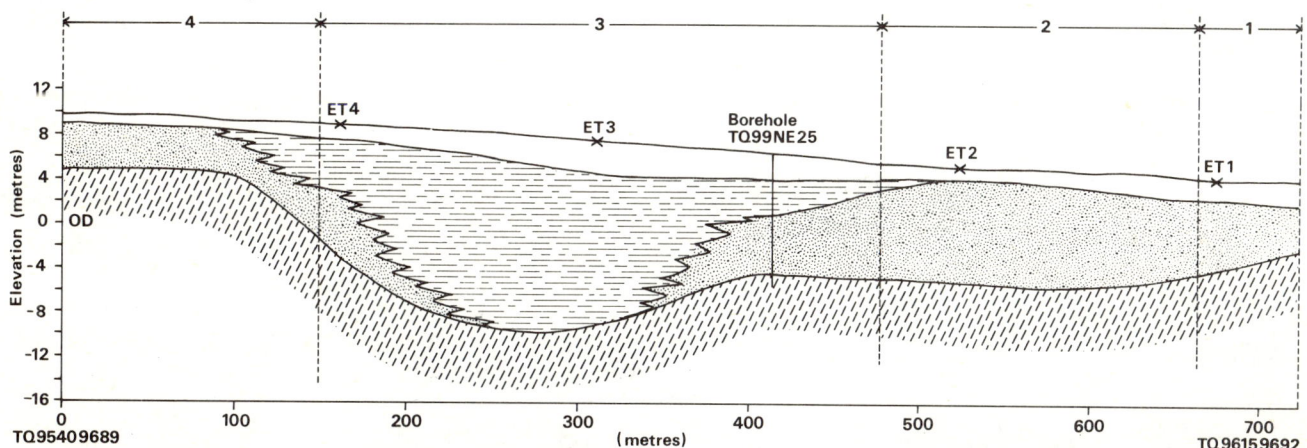

Fig. 10   Section across the Burnham Buried Channel

85

The traverse has been divided into four sections for ease of reference between the two diagrams.

Four vertical profiles using an expanding configuration were produced to enable a complete cross-section to be drawn. The centre point of each configuration lay along the line of the horizontal traverse at points E.T.1. [9613 9680], E.T.2. [9596 9680], E.T.3. [9576 9681] and E.T.4. [9560 9681]. The measured apparent resistivities Qa were plotted against half of the electrode separation on a bilogarithmic scale and the resultant curves compared with standard curves representing simulated geological conditions from which a quantitative interpretation was made. The results are summarised below where Q_1 Q_2 Q_3 and Q_4 are the apparent resistivity values in ohm metres for the first, second, third and fourth units respectively and where d_1 is the depth in metres of the base of unit 1, d_2 the depth of the base of unit 2 and d_3 the depth of the base of unit 3.

Although the use of geophysical methods was limited in this survey, it has been shown that when lateral variation in the geology at a locality is small, resistivity surveys may be used to supplement borehole information to aid the delimitation of terrace gravels beneath overburden and to determine the shapes and dimensions of channels.

Table 5. Results of expanding traverse near Dammer Wick Farm

	Q_1	Q_2	Q_3	Q_4	d_1	d_2	d_3
	ohm metres				depth in metres		
E.T.1	22	34	47	7.3	0.54	0.81	8.4
E.T.2	13	32.5	41	8.0	0.86	1.20	10.5
E.T.3	13	5.3	17	6.8	0.94	1.55	infinity
E.T.4	21.5	8.5	24.5	12.0	1.5	4.8	13

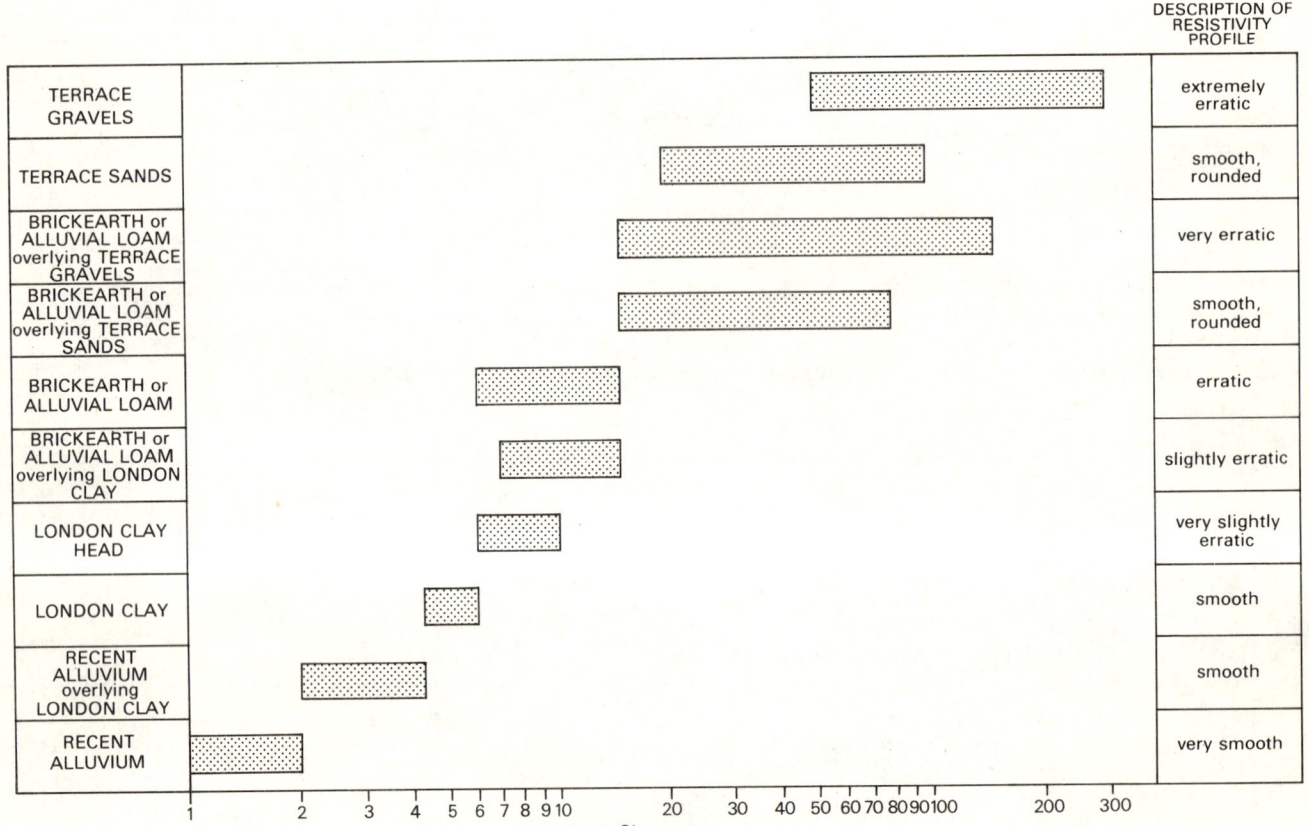

Fig. 11 Apparent resistivity ranges for certain lithologies and superficial deposits in south-east Essex

APPENDIX J: LIST OF QUARRIES ON THE DENGIE PENINSULA

Working Pits

Location	Grid Reference	Deposit worked
Asheldham	973 018	River Terrace and ? Glacial Gravels
Asheldham Chase, Asheldham	975 009	River Terrace and ? Glacial Gravels
Goldsand Road, Southminster	958 992	River Terrace and ? Glacial Gravels
Ratsborough, Southminster	951 985	River Terrace and ? Glacial Gravels
Curry, Nr. Bradwell	993 057	River Terrace and ? Glacial Gravels

Abandoned Pits

Location	Grid Reference	Deposit worked
Bradwell Hall, Nr. Bradwell	990 952	River Terrace and ? Glacial Gravels
Stow's Farm, Tillingham	985 030	River Terrace and ? Glacial Gravels

APPENDIX K: CONVERSION TABLE, METRES TO FEET (TO NEAREST 0.5 FT)

m	ft	m	ft	m	ft	m	ft	m	ft
0.1	0.5	6.1	20	12.1	39.5	18.1	59.5	24.1	79
0.2	0.5	6.2	20.5	12.2	40	18.2	59.5	24.2	79.5
0.3	1	6.3	20.5	12.3	40.5	18.3	60	24.3	79.5
0.4	1.5	6.4	21	12.4	40.5	18.4	60.5	24.4	80
0.5	1.5	6.5	21.5	12.5	41	18.5	60.5	24.5	80.5
0.6	2	6.6	21.5	12.6	41.5	18.6	61	24.6	80.5
0.7	2.5	6.7	22	12.7	41.5	18.7	61.5	24.7	81
0.8	2.5	6.8	22.5	12.8	42	18.8	61.5	24.8	81.5
0.9	3	6.9	22.5	12.9	42.5	18.9	62	24.9	81.5
1.0	3.5	7.0	23	13.0	42.5	19.0	62.5	25.0	82
1.1	3.5	7.1	23.5	13.1	43	19.1	62.5	25.1	82.5
1.2	4	7.2	23.5	13.2	43.5	19.2	63	25.2	82.5
1.3	4.5	7.3	24	13.3	43.5	19.3	63.5	25.3	83
1.4	4.5	7.4	24.5	13.4	44	19.4	63.5	25.4	83.5
1.5	5	7.5	24.5	13.5	44.5	19.5	64	25.5	83.5
1.6	5	7.6	25	13.6	44.5	19.6	64.5	25.6	84
1.7	5.5	7.7	25.5	13.7	45	19.7	64.5	25.7	84.5
1.8	6	7.8	25.5	13.8	45.5	19.8	65	25.8	84.5
1.9	6	7.9	26	13.9	45.5	19.9	65.5	25.9	85
2.0	6.5	8.0	26	14.0	46	20.0	65.5	26.0	85.5
2.1	7	8.1	26.5	14.1	46.5	20.1	66	26.1	85.5
2.2	7	8.2	27	14.2	46.5	20.2	66.5	26.2	86
2.3	7.5	8.3	27	14.3	47	20.3	66.5	26.3	86.5
2.4	8	8.4	27.5	14.4	47	20.4	67	26.4	86.5
2.5	8	8.5	28	14.5	47.5	20.5	67.5	26.5	87
2.6	8.5	8.6	28	14.6	48	20.6	67.5	26.6	87.5
2.7	9	8.7	28.5	14.7	48	20.7	68	26.7	87.5
2.8	9	8.8	29	14.8	48.5	20.8	68	26.8	88
2.9	9.5	8.9	29	14.9	49	20.9	68.5	26.9	88.5
3.0	10	9.0	29.5	15.0	49	21.0	69	27.0	88.5
3.1	10	9.1	30	15.1	49.5	21.1	69	27.1	89
3.2	10.5	9.2	30	15.2	50	21.2	69.5	27.2	89
3.3	11	9.3	30.5	15.3	50	21.3	70	27.3	89.5
3.4	11	9.4	31	15.4	50.5	21.4	70	27.4	90
3.5	11.5	9.5	31	15.5	51	21.5	70.5	27.5	90
3.6	12	9.6	31.5	15.6	51	21.6	71	27.6	90.5
3.7	12	9.7	32	15.7	51.5	21.7	71	27.7	91
3.8	12.5	9.8	32	15.8	52	21.8	71.5	27.8	91
3.9	13	9.9	32.5	15.9	52	21.9	72	27.9	91.5
4.0	13	10.0	33	16.0	52.5	22.0	72	28.0	92
4.1	13.5	10.1	33	16.1	53	22.1	72.5	28.1	92
4.2	14	10.2	33.5	16.2	53	22.2	73	28.2	92.5
4.3	14	10.3	34	16.3	53.5	22.3	73	28.3	93
4.4	14.5	10.4	34	16.4	54	22.4	73.5	28.4	93
4.5	15	10.5	34.5	16.5	54	22.5	74	28.5	93.5
4.6	15	10.6	35	16.6	54.5	22.6	74	28.6	94
4.7	15.5	10.7	35	16.7	55	22.7	74.5	28.7	94
4.8	15.5	10.8	35.5	16.8	55	22.8	75	28.8	94.5
4.9	16	10.9	36	16.9	55.5	22.9	75	28.9	95
5.0	16.5	11.0	36	17.0	56	23.0	75.5	29.0	95
5.1	17	11.1	36.5	17.1	56	23.1	76	29.1	95.5
5.2	17	11.2	36.5	17.2	56.5	23.2	76	29.2	96
5.3	17.5	11.3	37	17.3	57	23.3	76.5	29.3	96
5.4	17.5	11.4	37.5	17.4	57	23.4	77	29.4	96.5
5.5	18	11.5	37.5	17.5	57.5	23.5	77	29.5	97
5.6	18.5	11.6	38	17.6	57.5	23.6	77.5	29.6	97
5.7	18.5	11.7	38.5	17.7	58	23.7	78	29.7	97.5
5.8	19	11.8	38.5	17.8	58.5	23.8	78	29.8	98
5.9	19.5	11.9	39	17.9	58.5	23.9	78.5	29.9	98
6.0	19.5	12.0	39.5	18.0	59	24.0	78.5	30.0	98.5

REFERENCES

ALLEN, V.T. 1936. Terminology of medium-grained sediments. Rep. Natl. Res. Counc. Washington, 1935-36. App. 1, Rep. Comm. Sedimentation, pp. 18-47.

AMBROSE, J.D. 1973a. The sand and gravel resources of the country around Maldon, Essex: Description of 1:25 000 resource sheet TL 80. Rep. Inst. Geol. Sci., No. 73/1, 60 pp.

———— 1973b. The sand and gravel resources of the country around Layer Breton and Tolleshunt D'Arcy, Essex: Description of 1:25 000 resource sheet TL 91 and part of TL 90. Rep. Inst. Geol. Sci., No. 73/8, 34 pp.

ARCHER, A.A. 1969. Background and problems of an assessment of sand and gravel resources in the United Kingdom. Proc. 9th Commonw. Min. Metall. Congr., Vol. 2, Mining and Petroleum Geology. (London: Institution of Mining and Metallurgy), pp. 495-508.

———— 1970a. Standardisation of the size classification of naturally occurring particles. Géotechnique, Vol. 20, pp. 103-207.

———— 1970b. Making the most of metrication. Quarry Mgrs' J., Vol. 54, pp. 223-227.

ATTERBERG, A. 1905. Die rationelle Klassifikation der Sande und Kiese. Chem. Z., Vol. 29, pp. 195-198.

BRITISH STANDARD 1965. Aggregates for granolithic concrete floor finishes. Br. Stand., No. BS 1201, 8 pp.

———— 1967. Methods for sampling and testing of mineral aggregates, sand and fillers. Br. Stand., No. BS 812, 104 pp.

———— 1967. Methods of testing soils for civil engineering purposes. Br. Stand., No. BS 1377, 233 pp.

———— 1967. Specification for gravel aggregates for surface treatment (including surface dressings) on roads. Br. Stand., No. BS 1984, 8 pp.

BUREAU OF MINES AND GEOLOGICAL SURVEY. 1948. Mineral Resources of the United States (Bur. Mines and Geol. Surv.) (Washington, D.C.: Public Affairs Press), pp. 14-17.

DAVIES, M.C. and others. 1965. Records of wells in the area of New Series one-inch (geological) Epping (240), Chelmsford (241) and Brightlingsea (242) sheets. Wat. Supply Pap. Geol. Surv. G.B., Well Catalog. Ser. pp. 1-23.

GREENSMITH, J.T. and TUCKER, E.V. 1971. The effects of late Pleistocene and Holocene sea level changes in the vicinity of the River Crouch, east Essex. Proc. Geol. Assoc., Vol. 82, Part 3, pp. 301-321.

———— ———— 1973. Holocene transgressions and regressions on the Essex coast, outer Thames Estuary. Geol. Mijnb., Vol. 52, No. 4, pp. 193-202.

GRUHN, R., BRYAN, A.L. and MOSS, A.J. 1974. A contribution to Pleistocene chronology in south-east Essex, England. Quaternary Res., Vol. 4., pp. 53-71.

HARRIS, P.M., THURRELL, R.G., HEALING, R.A. and ARCHER, A.A. 1974. Aggregates in Britain. Proc. R. Soc., A 339, pp. 329-353.

HOLLYER, S.E. 1978. The sand and gravel resources of the country north and east of Southend-on-Sea, Essex: Description of parts of TQ 88/89, 98/99, TR 08/09. Miner. Assess. Rep. Inst. Geol. Sci. No.36, 212 pp.

LAKE, R.D., ELLISON, R.A., HOLLYER, S.E. and SIMMONS, M. 1977. Buried channel deposits in the south-east Essex area; their bearing on Pleistocene palaeogeography. Rep. Inst. Geol. Sci., No. 77/21, 13 pp.

LANE, E.W. and others. 1947. Report of the subcommittee on sediment terminology. Trans. Am. Geophys. Union, Vol. 28, pp. 936-938.

LEA, F.M. 1970. The Chemistry of Cement and Concrete. (Third Edition) (Edward Arnold (Publishers) Ltd.) p. 565.

MILNER, H.B. 1945. The natural history of gravel. Part 2. Distribution of gravel in England and Wales. Essex. Cem. Lime Gravel, Vol. 19, No. 12, pp. 429-438.

PETTIJOHN, F.J. 1957. Sedimentary Rocks (Second Edition) (London: Harper and Row).

THURRELL, R.G. 1971. The assessment of mineral resources with particular reference to sand and gravel. Quarry Mgrs' J., Vol. 55, pp. 19-25.

TWENHOFEL, W.H. 1937. Terminology of the fine-grained mechanical sediments. Rep. Natl. Res. Counc. Washington 1936-37. App. 1. Rep. Comm. Sedimentation, pp. 81-104.

UDDEN, J.A. 1914. Mechanical composition of clastic sediments. Bull. Geol. Soc. Am., Vol. 25, pp. 655-744.

WENTWORTH, C.K. 1922. A scale of grade and class terms for clastic sediments. J. Geol., Vol. 30, pp. 377–392.

————— 1935. The terminology of coarse sediments. Bull. Natl. Res. Counc. Washington, No. 98, pp. 225–246.

WILLMAN, H.B. 1942. Geology and mineral resources of the Marseilles, Ottawa and Streator quadrangles. Bull. Ill. State Geol. Surv., No. 66, pp. 343–344.

Dd 595761 K8
Printed in England for Her Majesty's Stationery Office by Commercial Colour Press, London

The following reports of the Institute relate particularly to bulk mineral resources

Reports of the Institute of Geological Sciences

Assessment of British Sand and Gravel Resources

1　The sand and gravel resources of the country south-east of Norwich, Norfolk: Resource sheet TG 20.　E. F. P. Nickless.
Report 71/20　ISBN 0 11 880216　£1.15

2　The sand and gravel resources of the country around Witham, Essex: Resource sheet TL 81.　H. J. E. Haggard.
Report 72/6　ISBN 0 11 880588 6　£1.20

3　The sand and gravel resources of the area south and west of Woodbridge, Suffolk: Resource sheet TM 24.　R. Allender and S. E. Hollyer.
Report 72/9　ISBN 0 11 880596 7　£1.70

4　The sand and gravel resources of the country around Maldon, Essex: Resource sheet TL 80.　J. D. Ambrose.
Report 73/1　ISBN 0 11 880600 9　£1.20

5　The sand and gravel resources of the country around Hethersett, Norfolk: Resource sheet TG 10.
E. F. P. Nickless.
Report 73/4　ISBN 0 11 880606 8　£1.60

6　The sand and gravel resources of the country around Terling, Essex: Resource sheet TL 71.　C. H. Eaton.
Report 73/5　ISBN 0 11 880608 4　£1.20

7　The sand and gravel resources of the country around Layer Breton and Tolleshunt D'Arcy, Essex: Resource sheet TL 91 and part of TL 90.　J. D. Ambrose.
Report 73/8　ISBN 0 11 990614 9　£1.30

8　The sand and gravel resources of the country around Shotley and Felixstowe, Suffolk: Resource sheet TM 23.
R. Allender and S. E. Hollyer.
Report 73/13　ISBN 0 11 880625 4　£1.60

9　The sand and gravel resources of the country around Attlebridge, Norfolk: Resource sheet TG 11.
E. F. P. Nickless.
Report 73/15　ISBN 0 11 880658 0　£1.85

10　The sand and gravel resources of the country west of Colchester, Essex: Resource sheet TL 92.　J. D. Ambrose.
Report 74/6　ISBN 0 11 880671 8　£1.45

11　The sand and gravel resources of the country around Tattingstone, Suffolk: Resource sheet TM 13.　S. E. Hollyer.
Report 74/9　ISBN 0 11 880675 0　£1.95

12　The sand and gravel resources of the country around Gerrards Cross, Buckinghamshire: Resource sheets SU 99, TQ 08 and TQ 09.　H. C. Squirrell.
Report 74/14　ISBN 0 11 880710 2　£2.20

Mineral Assessment Reports

13　The sand and gravel resources of the country east of Chelmsford, Essex: Resource sheet TL 70.　M. R. Clark.
ISBN 0 11 880744 7　£3.50

14　The sand and gravel resources of the country east of Colchester, Essex: Resource sheet TM 02.　J. D. Ambrose.
ISBN 0 11 880745 5　£3.25

15　The sand and gravel resources of the country around Newton on Trent, Lincolnshire: Resource sheet SK 87.
D. Price.
ISBN 0 11 880746 3　£3.00

16　The sand and gravel resources of the country around Braintree, Essex: Resource sheet TL 72.　M. R. Clark.
ISBN 0 11 880747 1　£3.50

17　The sand and gravel resources of the country around Besthorpe, Nottinghamshire: Resource sheet SK 86 and part of SK 76.　J. R. Gozzard.
ISBN 0 11 880748 X　£3.00

18　The sand and gravel resources of the Thames Valley, the country around Cricklade, Wiltshire: Resource sheets SU 09/19 and parts of SP 00/10.　P. R. Robson.
ISBN 0 11 880749 8　£3.00

19　The sand and gravel resources of the country south of Gainsborough, Lincolnshire: Resource sheet SK 88 and part of SK 78.　J. H. Lovell.
ISBN 0 11 880750 1　£2.50

20　The sand and gravel resources of the country east of Newark upon Trent, Nottinghamshire: Resource sheet SK 85
ISBN 0 11 880751 X　£2.75

21　The sand and gravel resources of the Thames and Kennet Valleys, the country around Pangbourne, Berkshire: Resource sheet SU 67.　H. C. Squirrell.
ISBN 0 11 880752 8　£3.25

22　The sand and gravel resources of the country north-west of Scunthorpe, Humberside: Resource sheet SE 81.
J. W. C. James.
ISBN 0 11 880753 6　£3.00

23　The sand and gravel resources of the Thames Valley, the country between Lechlade and Standlake: Resource sheet SP 30 and parts of SP 20, SU 29 and SU 39.　P. Robson.
ISBN 0 11 881252 1　£7.25

24　The sand and gravel resources of the country around Aldermaston, Berkshire: Parts of resource sheets SU 56 and SU 66.　H. C. Squirrell.
ISBN 0 11 881253 X　£5.00

25　The celestite resources of the area north-east of Bristol: Resource sheet ST 68 and parts of ST 59, 69, 79, 58, 78, 68 and 77.　E. F. P. Nickless, S. J. Booth and P. N. Mosley.
ISBN 0 11 881262 9　£5.00

26　The limestone and dolomite resources of the country around Monyash, Derbyshire: Resource sheet SK 16.
F. C. Cox and D. McC. Bridge.
ISBN 0 11 881263 7　£7.00

27　The sand and gravel resources of the country west and south of Lincoln, Lincolnshire: Resource sheets SK 95, SK 96 and SK 97.　I. Jackson.
ISBN 0 11 884003 7　£6.00

28　The sand and gravel resources of the country around Eynsham, Oxfordshire: Resource sheet SP 40 and part of SP 41.　W. J. R. Harries.
ISBN 0 11 884012 6　£3.00

29　The sand and gravel resources of the country south-west of Scunthorpe, Humberside: Resource sheet SE 80.
J. H. Lovell.
ISBN 0 11 884013 4　£3.50

30　Procedure for the assessment of limestone resources. F. C. Cox, D. McC. Bridge and J. H. Hull.
ISBN 0 11 884030 4　£1.25

31　The sand and gravel resources of the country west of Newark upon Trent, Nottinghamshire. Resource sheet SK 75.
D. Price and P. J. Rogers.
ISBN 0 11 884031 2　£3.50

32　The sand and gravel resources of the country around Sonning and Henley. Resource sheets SU 77 and SU 78.
H. C. Squirrell.
ISBN 0 11 884032 0　£5.25

33　The sand and gravel resources of the country north of Gainsborough. Resource sheet SK 89.　J. Gozzard and D. Price.
ISBN 0 11 884033 9　£4.50

34　The sand and gravel resources of the Dengie Peninsula, Essex: Resource sheet TL 90, etc.　M. B. Simmons.
ISBN 0 11 884081 9　£5.00

35　The sand and gravel resources of the country around Darvel: Sheets NS 53, 63, etc.　E. F. P. Nickless, A. M. Aitken and A. A. McMillan.
ISBN 0 11 884082 7　£7.00

36　The sand and gravel resources of the country around Southend-on-Sea, Essex: Resource sheets TQ 78/79 etc.
S. E. Hollyer and M. B. Simmons.
ISBN 0 11 884083 5　£7.50

37   The sand and gravel resources of the country around
Bawtry, South Yorkshire: Resource sheet SK 69.
A. R. Clayton.
ISBN 0 11 884053 3   £5.75

38   The sand and gravel resources of the country around
Abingdon, Oxfordshire: Resource sheets SU 49, 59 and SP 40,
50.   C. E. Corser.
ISBN 0 11 884084 3   £5.50

Reports of the Institute of Geological Sciences

Other Reports

69/9   Sand and gravel resources of the inner Moray
Firth.   A. L. Harris and J. D. Peacock.
ISBN 0 11 880106 6   35p

70/4   Sands and gravels of the southern counties of
Scotland.   G. A. Goodlet.
ISBN 0 11 880105 8   90p

72/8   The use and resources of moulding sand in Northern
Ireland.   R. A. Old.
ISBN 0 11 881594 0   30p

73/9   The superficial deposits of the Firth of Clyde and its sea
lochs.   C. E. Deegan, R. Kirby, I. Rae and R. Floyd.
ISBN 0 11 880617 3   95p

77/1   Sources of aggregate in Northern Ireland (2nd
edition).   I. B. Cameron.
ISBN 0 11 881279 3   70p

77/2   Sand and gravel resources of the Grampian Region.
J. D. Peacock and others.
ISBN 0 11 881282 3   80p

77/5   Sand and gravel resources of the Fife Region.
M. A. E. Browne.
ISBN 0 11 884004 5   60p

77/6   Sand and gravel resources of the Tayside Region.
I. B. Paterson.
ISBN 0 11 884008 8   £1.40

77/8   Sand and gravel resources of the Strathclyde
Region.   I. B. Cameron and others.
ISBN 0 11 884028 2   £2.50

77/9   Sand and gravel resources of the Central Region,
Scotland.   M. A. E. Browne.
ISBN 0 11 884016 9   £1.35

77/19   Sand and gravel resources of the Borders Region of
Scotland.   A. D. McAdam.
ISBN 0 11 884025 8   £1.00

77/22   Sand and gravel resources of the Dumfries and
Galloway Region of Scotland.   I. B. Cameron.
ISBN 0 11 884021 5   £1.20

78/1   Sand and gravel resources of the Lothian Region of
Scotland.   A. D. McAdam.
ISBN 0 11 884042 8   £1.00